Mechanical Behavior of Polymeric Materials: Recent Studies, 2nd Edition

Mechanical Behavior of Polymeric Materials: Recent Studies, 2nd Edition

Guest Editors

Emilia P. Collar
Jesús-María García-Martínez

MDPI Basel • Beijing • Wuhan • Barcelona • Belgrade • Novi Sad • Cluj • Manchester

Guest Editors
Emilia P. Collar
Institute of Polymer Science and Technology (ICTP)
Spanish National Research Council (CSIC)
Madrid
Spain

Jesús-María García-Martínez
Institute of Polymer Science and Technology (ICTP)
Spanish National Research Council (CSIC)
Madrid
Spain

Editorial Office
MDPI AG
Grosspeteranlage 5
4052 Basel, Switzerland

This is a reprint of the Special Issue, published open access by the journal *Polymers* (ISSN 2073-4360), freely accessible at: https://www.mdpi.com/journal/polymers/special_issues/L6W1RQ8R9P.

For citation purposes, cite each article independently as indicated on the article page online and as indicated below:

Lastname, A.A.; Lastname, B.B. Article Title. *Journal Name* **Year**, *Volume Number*, Page Range.

ISBN 978-3-7258-6956-5 (Hbk)
ISBN 978-3-7258-6957-2 (PDF)
https://doi.org/10.3390/books978-3-7258-6957-2

Contents

About the Editors

Emilia P. Collar

Emilia P. Collar holds a Ph.D. in Industrial Chemistry (U. Complutense, 1986). Since 1990, she has been a permanent staff member (Tenured Scientist) at the Consejo Superior de Investigaciones Científicas (CSIC), after two years (1986–88) as CSIC's postdoctoral fellow and one year (1989) as Chemical Engineering Assistant Teacher at the Universidad Complutense de Madrid. At the Polymer Science and Technology Institute (ICTP/CSIC), she works in the Polymer Engineering Group (GIP), founded in 1982 by Prof. O. Laguna, and has been its Head since 1999. Between 1990 and 2005, she supervised six doctoral and five postdoctoral theses across 15 research projects funded by public agencies, in addition to 18 research-funded contracts, and issued 53 technical reports for different companies. Furthermore, from 2001 to 2005, she held the positions of Technical Director of the Physical and Mechanical Properties Laboratory and Deputy Technical Director of the Thermal Properties Laboratory under the successful ISO 17025 Accreditation Project for the CSIC/ICTP's Laboratories, ACiTP, commanded by the ICTP's Head. She is the author and co-author of more than 20 book chapters, 2 currently active industrial patents on polymer recycling, and more than 150 papers, mainly in SCI Journals. Her research focuses on polymers and the environment within the general framework of heterogeneous polymer-based materials. From 2006 to date, she has participated in four public Spanish-funded research projects and one EU project under its 7th Framework Program.

Jesús-María García-Martínez

Jesús-María García-Martínez holds a Ph.D. in Chemistry (Chemical Engineering) from the Universidad Complutense de Madrid (1995), two M.Sc. degrees (Industrial Chemistry and Polymer Science and Technology), and has completed more than 100 highly specialized courses. He serves as a Senior Scientist at the Institute of Polymer Science and Technology (ICTP) of the Spanish National Research Council (CSIC). Since 1992, within the Polymer Engineering Group (GIP), he has co-authored more than 200 scientific and/or technical works on topics related to polymer engineering, chemical modification of polymers, heterogeneous polymer-based materials, interphases and interfaces, polymer composites, polyblends and alloys, organic–inorganic hybrid materials, polymer recycling, quality, and so on. Furthermore, he has participated in 32 research and industrial projects (national and international programs) and is a co-author of 1 currently active industrial patent on polymer recycling. From 2000–2005, he also served as the Quality Director for the ICTP (CSIC) ISO 17025 Accreditation Project for the ICTP Laboratories. Additionally, Dr. García-Martínez is actively reviewing tasks for WOS and SCOPUS-indexed journals, with more than 600 reports in the last few years. He has been awarded the Publons Reviewer Award (2018, 2019) twice, as well as the POLYMERS Outstanding Reviewer Award (2019). He is an Editorial Board Member of polymer science-related journals and the Editor of more than 150 scientific articles. He has been a Guest Editor of ten Special Issues published by Q1 journals. Since 2016, he has been the Head of the Department of Chemistry and Properties of Polymer Materials within the ICTP (CSIC).

Preface

This Special Issue addresses the complex and evolving field of polymer science and technology, with particular emphasis on the factors that influence the properties of polymer-based materials. Such knowledge is not solely of academic interest; it is a critical requirement in both polymer science and engineering. Thus, the articles presented in this new Special Issue exemplify both current and emerging trends in the study of the mechanical properties of polymeric materials. Collectively, these nine contributions provide valuable insights and serve as a foundation for future research in this dynamic scientific field.

Emilia P. Collar and Jesús-María García-Martínez
Guest Editors

polymers

MDPI

Editorial

New Features on Mechanical Behavior of Polymeric Materials

Emilia P. Collar * and Jesús-María García-Martínez *

Polymer Engineering Group (GIP), Polymer Science and Technology Institute (ICTP), Spanish National Research Council (CSIC), C/Juan de la Cierva, 3, 28006 Madrid, Spain
* Correspondence: ecollar@ictp.csic.es (E.P.C.); jesus.maria@ictp.csic.es (J.-M.G.-M.)

This Special Issue, devoted to new features on "Mechanical Behavior of Polymeric Materials", includes many exciting works related to this frontrunner polymer Research and Development area. The topic's fundamentals underscore their significance and address a dynamic, rapidly advancing area of polymer science. Specifically, it examines the diverse factors that influence the properties of polymer-based materials. It is essential to provide readers with foundational information on the concept of failure in polymer-based materials, especially as it relates to their mechanical properties. From a fundamental perspective, a fracture in a material occurs when the forces binding its constituent atoms are overcome. However, the specific ways in which these atoms are bonded in molecules and the resulting diversity of supramolecular aggregates lead to complex questions and contribute to the heterogeneous nature of these materials [1,2]. In evaluating material strength, it has been well established since the early studies by Griffith [3] that fracture consistently initiates in the weakest region of a material. However, the pursuit of empirical and semi-empirical approaches, grounded in the results of rigorously designed mechanical testing procedures, remains an open research question, particularly when these procedures are limited to material or part performance. These considerations are particularly significant for heterogeneous materials composed entirely or partially of organic polymers. Since the early stages of materials research [4–6], it has been established that the stress–strain relationship in such materials is nonlinear, as it is strongly influenced by time- and temperature-dependent effects on the applied external force, as well as by morphological changes within the material's bulk resulting from environmental conditions [7–9]. This underscores the continued importance of research in this field to fulfill key scientific requirements and ensure reproducible, reliable results. The development of inter-laboratory protocols and international standardized procedures is crucial for providing robust estimates of mechanical parameters [10–15]. These values should be suitable for incorporation into comprehensive databases to support polymeric part and material design software [16], enabling accurate forecasting of the mechanical behavior of polymeric materials once they reach the solid-state following processing steps that determine the resulting morphologies of the organic fractions in these materials [17,18]. Recent findings demonstrate that the interaction between mechanical energy and matter varies significantly across different environments and conditions. This, combined with rigorous analysis of material responses, remains essential for developing mathematical models to optimize material design within a comprehensive sustainability framework, ultimately aiming to prevent catastrophic failure scenarios.

To mention that a substantial number of manuscripts (15 articles) were submitted for consideration in this Special Issue; however, only a select group (eight documents) was published after the rigorous review process of *Polymers*, yielding a 47% acceptance rate. The

 https://doi.org/10.3390/polym18020266

articles compiled in this volume are fully aligned with the research philosophies outlined above. It is noteworthy to note that the purpose of this Editorial is not to discuss each article in detail, but rather to encourage readers to engage with them thoroughly. Therefore, since this editorial aims not to elaborate on each text but to encourage the reader to browse them in depth, these contributions have been briefly described below. For such purposes, a short note on each one has been reported to awaken interest in each of the contributions to this exciting Special Issue of *Polymers*, rather than providing an exhaustive description of each of them.

In this way, the article by Sun et al. [19] introduces a novel approach to enhance the reinforcement between binders and high-solid fillers in propellant formulations by grafting bonding groups onto the binder to form a neutral polymeric structure. For such a purpose, the authors employ a glycidyl azide polyol energetic thermoplastic elastomer binder with a –CN bonding group (GAP–ETPE), which was synthesized and tested with RDX-based model propellants. Furthermore, the mechanical analysis showed improved tensile strength (6.43 MPa) and strain (32.1%). Additionally, dynamic mechanical analysis indicated that increasing the RDX content raised the glass transition temperature (Tg) and the storage modulus. And finally, thermal analysis revealed four decomposition stages, enabling the authors to establish a thermal decomposition equation. These findings provide a promising method for improving mechanical properties and understanding thermal decomposition behavior, offering technical support for propellant combustion studies.

The study by Trindade and colleagues [20] explores the use of additive manufacturing, particularly fused deposition modeling, to produce customized orthoses that improve patient comfort and quality of life. For such purposes, a series of nine polymeric materials was evaluated through compressive, flexural, and tensile tests in both horizontal and vertical print orientations. The authors concluded that polycarbonate, polylactic acid, and ULTEM™ 1010 demonstrated superior mechanical properties and consistent performance across orientations, making them ideal for orthotic applications. Additionally, a finite element model of an ankle–foot orthosis was developed to simulate stress, strain, and deformation under static conditions. Results highlight ULTEM™ 1010 as the most durable and high-performing material, providing guidance for optimizing orthotic fabrication.

Costa and coworkers [21] examine the notch effect on semicrystalline PVDF using U- and V-notch geometries of varying depths, performing tensile tests at 23 °C and combining Digital Image Correlation (DIC) and Finite Element Analysis (FEA). Both unnotched and notched specimens were analyzed to compare global mechanical curves and local strain maps. Thus, the results showed that notch geometry and depth significantly reduce load and displacement compared to unnotched samples, with strain maps confirming localized strain concentration near notch tips. Additionally, FEA demonstrated strong agreement with experimental data globally and reasonable accuracy locally within 0.5 mm of the notch region. Overall, DIC and FEA proved effective for evaluating notch behavior in PVDF used as pressure sheaths.

The work by Abdellah et al. [22] focused on composite materials, valued for their high strength-to-weight ratio, which are widely used in the aerospace, automotive, and shipbuilding industries to reduce energy consumption. They pay attention to the damage behavior, which changes significantly when stress discontinuities, such as holes, are introduced into the composite. This study investigates the effect of multiple holes on carbon fiber composites using a progressive damage model and finite element analysis (FEA). In this way, two-hole configurations were analyzed along longitudinal and transverse directions relative to the load. Results show that additional holes can act as stress-relief features, reducing stress by up to 17% when aligned longitudinally. Thus, a cohesive zone model was applied to develop a simplified analytical method for predicting the nominal

https://doi.org/10.3390/polym18020266

strength of multi-hole laminates based on unnotched plate properties. They conclude that the model closely matched experimental data, and design tables were provided to support material selection and structural optimization.

Mohammed, Abdellah and colleagues [23] investigate the prediction of mode I fracture energy in Graphite-reinforced composite laminates. These types of materials are widely used due to their versatility and high-performance properties, but have the problem that delamination remains the most critical failure mode. Thus, the subject of this study is to predict mode I interlaminar fracture energy using the Virtual Crack Closure Technique (VCCT) integrated with finite element modeling (FEM) on double cantilever beam (DCB) specimens. Additionally, the authors developed a simple analytical model based on material strength and stiffness to calculate the critical fracture energy. They conclude that both approaches showed strong agreement with experimental results, yielding an error margin of 5%, which confirms their accuracy and reliability for evaluating delamination behavior under service conditions.

The article by Morokov et al. [24] investigates the effect of adding single-walled carbon nanotubes (CNTs) to the polymer matrix of carbon fiber laminates to enhance strength and resistance to mechanical loads. In this context, the development of impact damage was analyzed using high-resolution ultrasound imaging for laminates with CNT concentrations ranging from 0 to 0.5 wt%. Further, their results show that CNT addition reduces damage in the upper and lower layers but increases damage in the middle plies. These findings were discussed alongside impact history data, providing insights into the role of CNTs in damage propagation within composite laminates.

The investigation by Zhang et al. [25] evaluates the mechanical properties of ABS parts produced via fused deposition modeling (FDM) using a combination of experimental and numerical methods. In this way, a series of ABS specimens underwent tensile testing on a universal testing machine, while finite element analysis (FEA) in ANSYS 2021 simulated stress and deformation behavior under varying conditions, including pre-stretching and temperature gradients. Experimental results showed a maximum tensile force of 7.3 kN, upper yield force of 3.7 kN, and lower yield force of 3.2 kN, indicating high strength and toughness. Additionally, non-proportional elongation reached 6%, and the performance enhancement factor compared to traditional manufacturing was 1.1, aligning with reinforced ABS standards. Finally, the FEA results validated experimental findings, confirming 15 mm of plastic deformation before fracture, consistent with ABS's ductile nature.

Finally, the research by Amaro et al. [26] focused on soft biological tissues that exhibit highly nonlinear and anisotropic mechanical behavior, posing challenges for replicating these properties in engineered materials for biomedical applications such as surgical simulation and device testing. This study presents a framework for reproducing the nonlinear stress–strain response of soft tissue using 3D-printed models. For such purposes, two polymers—thermoplastic polyurethane (TPU) and thermoplastic elastomer (TPE)—were selected for their tunable hardness and elasticity. The authors conducted a parametric study to examine the effects of Shore A hardness (60 A–100 A), infill density (0–100%), and shell number (0–2) on tensile performance. Additionally, mechanical testing provided stress–strain curves to evaluate structural reliability and functional behavior. In conclusion, these findings demonstrate the potential of additive manufacturing for anatomical reproduction and for accurate replication of mechanical properties in soft tissue models.

In summary, as Guest Editors of this Special Issue, and in light of the contributions to this Special Issue, we can affirm that the topic "Mechanical Behavior of Polymeric Materials" constitutes a fundamental framework within Polymer Science and Technology, both currently and in the foreseeable future. Accordingly, a third edition on this topic,

 https://doi.org/10.3390/polym18020266

scheduled for publication in 2026 in *Polymers*, is underway and now open for submissions. We welcome quality contributions.

Conflicts of Interest: The authors declare no conflicts of interest.

References

1. Collar, E.P.; García-Martínez, J.-M. Mechanical Behavior of Polymeric Materials: Recent Studies. *Polymers* **2024**, *16*, 2821. [CrossRef] [PubMed]
2. Collar, E.P.; García-Martínez, J.-M. (Eds.) *Mechanical Behavior of Polymeric Materials: Recent Studies*; MDPI Books: Basel, Switzerland, 2023; ISBN 978-3-7258-2524-0, 978-3-7258-2523-3.
3. Griffith, A.A. The Phenomena of Rupture and Flow in Solids. *Philos. Trans. R. Soc. Lond.* **1921**, *221*, 163–198. [CrossRef]
4. Peterlin, A. *Plastic Deformation of Polymers*; M. Dekker: New York, NY, USA, 1971.
5. Flory, P.J. *Principles of Polymer Chemistry*; Cornell University Press: New York, NY, USA, 1953.
6. Morawetz, H. *Polymers: The Origins and Growth of a Science*; John Wiley & Sons: New York, NY, USA, 1985.
7. Gugumus, F. The Use of Accelerated Tests in the Evaluation of Antioxidants and Light Stabilizers. In *Developments in Polymer Stabilisation-8*; Scott, G., Ed.; Springer: Dordrecht, The Netherlands, 1987.
8. Van Krevelen, D.W.; Nijenhuis, K.T. *Properties of Polymers: Their Correlation with Chemical Structure; Their Numerical Estimation and Prediction from Additive Group Contributions*; Elsevier: Amsterdam, The Netherlands, 2009.
9. Mascia, L. *The Role of Additives in Plastics*; Halsted Press: New York, NY, USA, 1975.
10. Mascia, L. *Thermoplastics: Materials Engineering*, 2nd ed.; Elsevier: New York, NY, USA, 1989; Chapter 4.
11. Cogswell, E.N. *Thermoplastic Aromatic Polymer Composites*; Butterworth-Heinemann: Oxford, UK, 1992.
12. Al Malaika, S.; Golovoy, A.; Wilkie, C.A. *Chemistry and Technology of Polymer Additives*; Blackwell Science: Oxford, UK, 1999.
13. Katz, H.S.; Milewski, J.V. *Handbook of Fillers and Reinforcements for Plastics*; Van Nostrand Reinhold: New York, NY, USA, 1978.
14. Roulin-Moloney, A.C. (Ed.) *Fractography and Failure Mechanisms of Polymers and Composites*; Kluwer Academic Publishers: Dordrecht, The Netherlands, 1989.
15. Carlsson, L.A.; Adams, D.F.; Pipes, R.B. *Experimental Characterization of Advanced Composite Materials*, 4th ed.; CRC Press: Boca Raton, Fl, USA, 2014.
16. Japan Association for Chemical Innovation (Ed.) *Computer Simulation of Polymeric Materials: Applications of the OCTA System*; Springer: Singapore, 2016.
17. Tadmor, Z.; Gogos, C.G. *Principles of Polymer Processing*, 2nd ed.; Wiley: New York, NY, USA, 1979.
18. Han, C.D. *Rheology in Polymer Processing*; Academic Press: Cambridge, MA, USA, 1976.
19. Sun, Q.; Yang, X.-M.; Yin, G.-Z. Mechanical Properties and Thermal Decomposition Mechanism of Glycidyl Azide Polyol Energetic Thermoplastic Elastomer Binder with RDX Composite. *Polymers* **2024**, *16*, 2626. [CrossRef] [PubMed]
20. Trindade, D.; Habiba, R.; Fernandes, C.; Costa, A.A.; Silva, R.; Alves, N.; Martins, R.; Malça, C.; Branco, R.; Moura, C. Material Performance Evaluation for Customized Orthoses: Compression, Flexural, and Tensile Tests Combined with Finite Element Analysis. *Polymers* **2024**, *16*, 2553. [CrossRef] [PubMed]
21. Pereira, I.C.S.; de Sousa, J.R.M.; Costa, C.A. Evaluation of U-Notch and V-Notch Geometries on the Mechanical Behavior of PVDF: The DIC Technique and FEA Approach. *Polymers* **2024**, *16*, 2906. [CrossRef] [PubMed]
22. Alharthi, H.; Abdellah, M.Y. Stress Analysis and Strength Prediction of Carbon Fiber Composite Laminates with Multiple Holes Using Cohesive Zone Models. *Polymers* **2025**, *17*, 124. [CrossRef] [PubMed]
23. Mohamed, A.F.; Abdellah, M.Y.; Hassan, M.K.; Backar, A.H. Advanced Prediction and Analysis of Delamination Failure in Graphite-Reinforced Epoxy Composites Using VCCT-Based Finite Element Modelling Techniques. *Polymers* **2025**, *17*, 771. [CrossRef] [PubMed]
24. Morokov, E.; Shershak, P.; Burkov, M.; Eremin, A.; Popkova, E.; Yakovlev, N.; Zhiltsova, I. Influence of CNT Filler in Polymer Matrix on Impact Damage Propagation in the Volume of Carbon Fiber Laminates. *Polymers* **2025**, *17*, 891. [CrossRef] [PubMed]
25. Li, Y.; Zhu, P.; Zhang, D. Experimental and Numerical Investigation of the Mechanical Properties of ABS Parts Fabricated via Fused Deposition Modeling. *Polymers* **2025**, *17*, 1957. [CrossRef] [PubMed]
26. Valvez, S.; Oliveira-Santos, M.; Gonçalves, L.; Piedade, A.P.; Amaro, A.M. Additive Manufacturing to Mimic the Nonlinear Mechanical Behavior of Cardiac Soft Tissue. *Polymers* **2025**, *17*, 2949. [CrossRef] [PubMed]

https://doi.org/10.3390/polym18020266

polymers

MDPI

Article

Mechanical Properties and Thermal Decomposition Mechanism of Glycidyl Azide Polyol Energetic Thermoplastic Elastomer Binder with RDX Composite

Qili Sun [1,*], Xiao-Mei Yang [2,*] and Guang-Zhong Yin [3]

1 School of Materials Science and Engineering, Nanjing University of Aeronautics and Astronautics, Nanjing 210016, China
2 Faculty of Design, Innovation and Technology, University of Design, Innovation and Technology (UDIT), Av. Alfonso XIII, 97, 28016 Madrid, Spain
3 Escuela Politécnica Superior, Universidad Francisco de Vitoria, Ctra. Pozuelo-Majadahonda Km 1.800, 28223 Madrid, Spain; amos.guangzhong@ufv.es
* Correspondence: sunqili2023@nuaa.edu.cn (Q.S.); xiaomei.yang@udit.es (X.-M.Y.)

Abstract: To improve the reinforcement effect between a binder and high solid filler in a propellant formula, grafting the bonding group into the binder to form a neutral polymeric is a practically novel approach to improving the interface properties of the propellant. In this work, a glycidyl azide polyol energetic thermoplastic elastomer binder with a –CN bonding group (GAP–ETPE) was synthesized, and the mechanical and thermal decomposition mechanism of GAP–ETPE with Hexogeon (RDX) model propellants were studied. The stress–strain results indicated that the tensile strength and strain of GAP–ETPE/RDX model propellants were 6.43 MPa and 32.1%, respectively. DMA data showed that the storage modulus (E′) of the GAP–ETPE/RDX model propellants could increase the glass transition temperature (*Tg*) values, those were shifted to higher temperature with the increase in filler RDX percentages. TG/DTG showed the four decomposition stages of the decomposition process of the GAP–ETPE/RDX model propellants, and the thermal decomposition equation was constructed. These efforts provide a novel method to improve GAP–ETPE/RDX propellants mechanical property, and the thermal decomposition behavior of GAP–ETPE/RDX propellants also provided technical support for the study of propellant combustion characteristics.

Keywords: bonding functional binder; energetic thermoplastic elastomer; polymer composites; mechanical properties; thermal decomposition mechanism

1. Introduction

Solid propellants were applied to medium-sized caliber rockets, and they were mainly composed of binders, high-energy fuels, and other additives, which are divided into thermosetting propellants and thermoplastic propellants. Thermosetting propellants require short-term casting and shaping after adding a curing agent, and they cannot be reused due to their crosslinking structure. In contrast, thermoplastic propellants have the advantages of emergency readiness, repeatable processing, and reuse, which has roused the attention of countries around the world [1–4]. Beaupre synthesized a glycidyl azide polyol energetic thermoplastic elastomer (GAP–ETPE) binder [5], and found that this binder has the advantages of being more thermally stable and insensitive to mechanical stimuli than expensive crystalline poly (bis(azidomethyl)oxetane (polyBAMO) [6,7]. Thus, the GAP–ETPE binder has attracted many researchers investigating its mechanical properties [8–10], rheological properties [11], and thermal safety performance [12], which shows good application prospects for the propellant. Furthermore, scientists often increase the specific impulse of propellants by filling them with high-proportion and high-energy explosive cyclotrimethylene trinitramine (RDX). However, it easily suffers from weak interface properties between

GAP–ETPE binder and filled RDX and causes unstable burning properties, even potential explosion and accidents when GAP–ETPE/RDX composite propellant undergoes high thrust during work [13,14].

To improve the interfacial properties of solid composite propellant, Zhang designed and synthesized a thermoplastic binder with bonding function, based on Kim's idea of a neutral polymeric bonding function binder, which can effectively resolve weak interfacial properties for propellants [15,16]; the results show that the tensile strength of the GAP–ETPE/RDX composite increases with the increase in the number of functional groups.

Regrettably, they did not further investigate the mechanism of interfacial enhanced mechanism. Furthermore, some researchers considered the decomposition process of the GAP–ETPE/RDX; however, they did not study how the filler RDX's decomposition influences the other component (GAP–ETPE binder). It is necessary to study the thermal decomposition behavior of propellant.

In this study, the GAP–ETPE binder with a –CN-bonding functional group was synthesized, and the interfacial enhanced mechanism between binder and RDX was proposed. Then, the mechanical properties and decomposition behaviors of the GAP–ETPE/RDX propellants were investigated by static mechanics analysis, dynamic mechanical analysis (DMA), TG/DTG, and TG/FTIR.

2. Experiments

2.1. Materials

Cyclotrimethylene trinitramine (RDX): Yingguang Chemical Company in Gansu, China, RDX diameter, average particle size of 40 μm; its structures are shown in Figure 1.

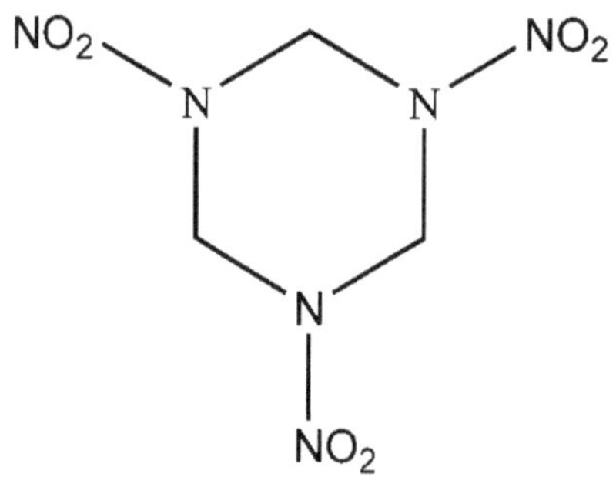

Figure 1. Structures of RDX.

The GAP–ETPE binder with the –CN group. The GAP–ETPE with the –CN group was synthesized based on the previous work [8], and the synthesis process was illustrated in Figure 2. The number-average molecular weight (M_n) was about 30,500 g·mol^{-1}, and the polydispersity index (PDI; M_w/M_n) of the GAP–ETPEs was 2.06; the ratio of the hard segment (with –CN groups) and soft segment of the GAP–ETPE binder was 3/7.

Preparation of the GAP–ETPE/RDX model propellant [17]: First, the GAP–ETPE was dissolved in tetrahydrofuran to form the GAP–ETPE polymer solution; afterwards, the RDX powder was added to the GAP–ETPE polymer solution at various weight percentages: 10, 30, 50, 70, and 80 and stirred for 0.5 h. The solvent was evaporated and dried in a vacuum oven at 70 °C for 6 h. Finally, the mixture was mixed in an open mill followed vulcanization molding. These samples were named GAP–ETPE/RDX-10, GAP–ETPE/RDX-30, GAP–ETPE/RDX-50, GAP–ETPE/RDX-70, GAP–ETPE/RDX-80, which correspond to RDX filler percentage of 10, 30, 50, 70, and 80, respectively.

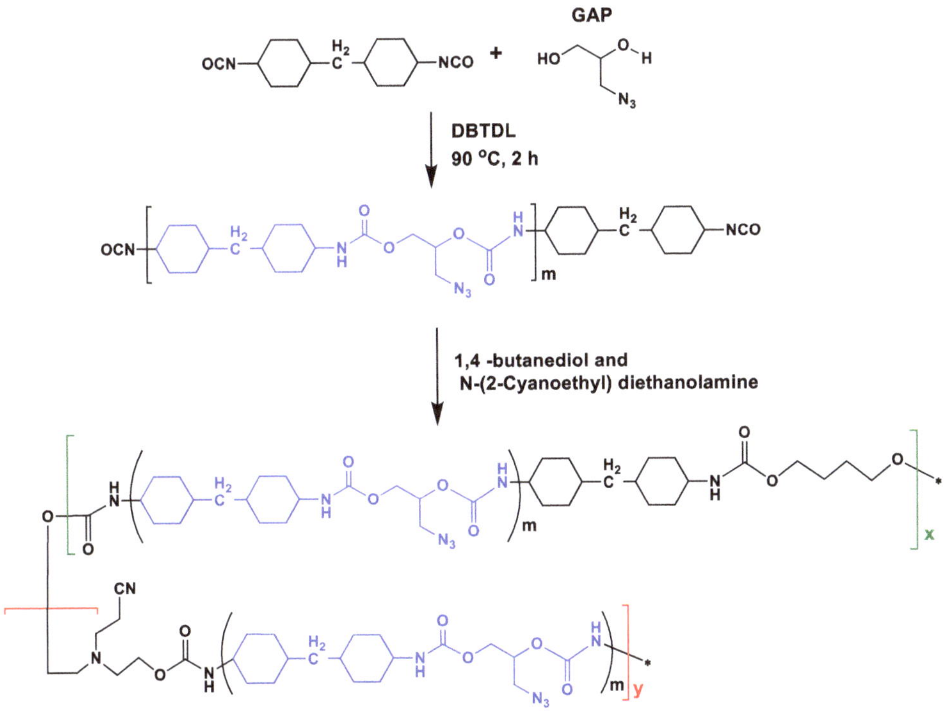

Figure 2. The synthesis process of the GAP–ETPE binder.

2.2. Measurements

***Tensile test*:** The stress–strain test was measured using a tensile testing machine (AGS-J, Shimadzu Co., Ltd., Kyoto, Japan) at a constant strain rate of 100 mm/min at room temperature. The dimensions of the dumbbell-shaped specimen samples were in 20 mm × 4 mm × 2 mm (test standards: GB/T 528-1998 [18]).

***Dynamic Mechanical Analyzers (DMA)*:** The dynamic thermomechanical properties were measured with a DMA (DMA/SDTA861e, METTLER TOLEDO Co., Ltd., Zurich, Switzerland). All samples were tested in thin film tension mode at an oscillating strain of 5 μm with the frequency of 1 Hz. The experiments were performed at a heating rate of 3 °C/min from −80 °C to 100 °C under a nitrogen atmosphere. The glass transition temperature (Tg) was determined as the temperature at the maximum of the tan δ vs. temperature curve.

***Scanning electron microscope (SEM)*:** Fractured surface morphology of GAP–ETPE/RDX model propellant was studied by Hitachi S4800 (Tungsten source, Hitachi High-Technologies Corp., Tokyo, Japan) Dynamic Mechanical Analyzers scanning electron microscope (SEM) at an accelerating voltage of 15 kV. The fracture surface of all samples was not coated.

***Thermogravimetric Analysis (TGA)*:** The decomposition behavior was observed on a METTLER TGA/DSC1/1600 Instrument simultaneous Differential Thermal Analysis (DTA)-TGA((TGA/DSC1/1600, METTLER TOLEDO Co., Ltd., Zurich, Switzerland)). Measurements were usually made with approximately 3 mg of sample using alumina cups. The samples were heated up to 600 °C at a heating rate of 3 °C/min, and a gas flow rate of 40 mL/min of high purity helium was circulated in the cell. Aluminum, silver, and zinc were utilized to calibrate the apparatus. We defined T5% as the onset decomposition temperature of composites, which means 5% weight loss, and T_{max} as the peak temperature of DTG, which means the highest weight loss rate.

TGA/FTIR: Thermal decomposition mechanism of the GAP–ETPE/RDX was performed by a TG/FTIR. A thermogravimetric analyzer, model TGA/DSC1SF/417-2 (Mettler-Toledo), was used over the range 30–600 °C at a rate of 10 K/min in an argon atmosphere (40 mL/min). A FTIR spectrophotometer, model Nicolet iS10 (Thermo Fisher Scientific, Massachusetts, USA), was linked to the thermogravimetric analyzer to measure the gas products. The FTIR transferring line was heated to 220 °C, and FTIR spectra were recorded from 500 to 4000 cm^{-1} with a resolution of 4 cm^{-1}.

3. Results and Discussion

3.1. Dynamic Mechanical Analysis

DMA is selected for characterizing the dynamic mechanical properties of the pure GAP–ETPE and its composites by filling RDX. Figure 3 shows the storage modulus (E′) curves of the pure GAP–ETPE and GAP–ETPE/RDX model propellant with different RDX percentages in the temperature range of −80 °C to 120 °C. It was observed that the E′ value presents a significant enhancement with increasing RDX contents. Typically, the E′ values of pure GAP–ETPE were about 7280 MPa below −40 °C; however, this value increased up to 17,000 MPa after adding RDX to the GAP–ETPE binder (GAP–ETPE/RDX-80). The great improvement in E′ values may be attributed to good matrix–filler interactions because of the induced effect of –CN and $–NO_2$; the induced effect model was constructed and will be discussed in Section 3.3.

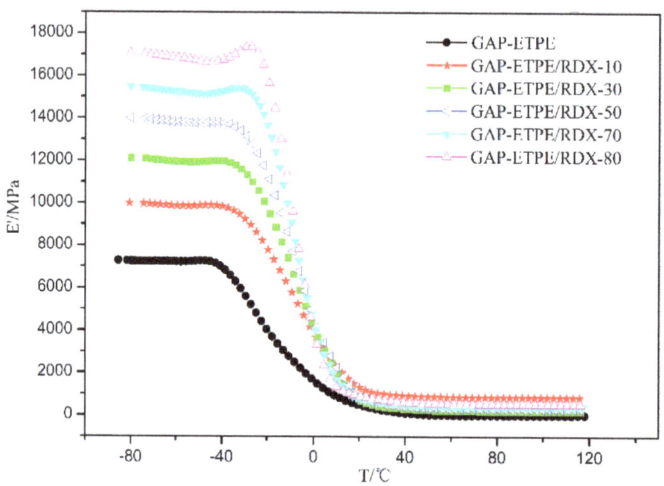

Figure 3. The E′ of samples as a function of temperature.

The damping factor (*tan δ*) as a function of temperature was shown in Figure 4. The temperature located on peak value can be considered as the glass transition temperature (*Tg*) of GAP–ETPE because of α relaxation corresponding to the glass transition of the polymer composite materials [19], and the related data were summarized in Table 1. From Table 1, it can be seen that the *Tg* of all the GAP–ETPE/RDX model propellants was higher than that of pure GAP–ETPE; meanwhile, the *Tg* value shifted to higher temperature (from −30.1 °C to −26.1 °C) with the RDX percentages increasing, which was attributed to the added RDX particles restricting the mobility of the chain segment of GAP–ETPE [20]. In addition, the pure GAP–ETPE binder showed a narrow *tan δ* peak, while the GAP–ETPE/RDX model propellant clearly showed a broad *tan δ* peak, which indicated that the relaxation process occurred slower than that of the reference sample due to matrix–particle interaction.

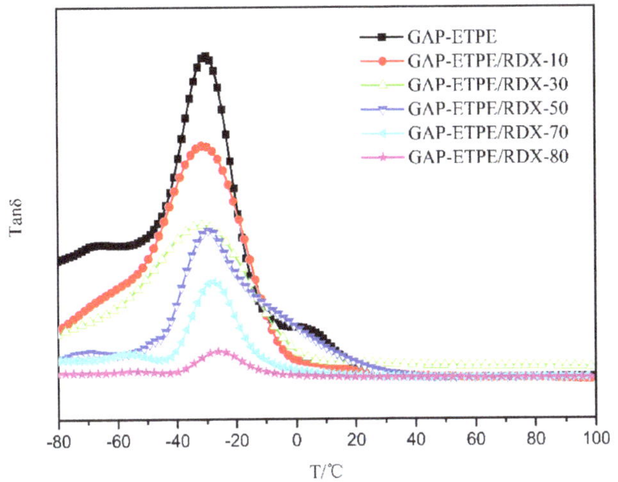

Figure 4. Tan δ of samples as a function of temperature.

Table 1. The *Tg* value of GAP–ETPE/RDX.

Samples	RDX/%	*Tg*/°C
GAP–ETPE	0	−31.6
GAP–ETPE/RDX-10	10	−30.1
GAP–ETPE/RDX-30	30	−28.6
GAP–ETPE/RDX-50	50	−27.2
GAP–ETPE/RDX-70	70	−26.4
GAP–ETPE/RDX-80	80	−26.1

3.2. Tensile Mechanical Properties Analysis

The tensile stress–strain curve of GAP–ETPE/RDX model propellants was shown in Figure 5. It can be observed that the GAP–ETPE/RDX-70 composite propellant exhibited the highest tensile strength. The enhancement of the mechanical properties may be due to good dispersion of GAP–ETPE composite materials filling low percentages RDX in the previous section, as well as good interfacial interaction between RDX and GAP–ETPE with –CN group matrix [21,22]. However, the tensile strength values decreased when the RDX percentages surpassed 70%, which may be associated with inevitable aggregation of the RDX powders at higher filler percentages. Some aggregations can be formed, thus leading to the presence of stress concentrators, and as a consequence, the mechanical properties decrease. On the other hand, the elongation at break point had decreased with the increase in RDX filler, which may be attributed to deformation of GAP–ETPE chains segment and micro-cracks appearing with high RDX percentages. These mechanical data showed that the GAP–ETPE-based propellant with –CN group could be a hopeful application to meet the high performance requirement of the propellant. The typical parameters are listed in Table 2.

In order to confirm the above supposition, the tensile fracture surface morphology of the distribution of RDX in the matrix was examined. Figure 6 shows that the surface micrographs of the GAP–ETPE/RDX model propellants with RDX-50, 70, and 80% correspond to Figure 6a, Figure 6b, and Figure 6c, respectively. As it can be seen from Figure 6, when the filled RDX ratios was less than 70%, there were very few aggregations of RDX grain surface micrographs structures observed on the GAP–ETPE/RDX model propellant. However, as the filler ratios increases to 80%, more aggregations of RDX grain spread on the GAP–ETPE/RDX model propellant surface. These results demonstrate that the aggregations of RDX grain influenced the tensile strength of the GAP–ETPE/RDX model propellant. Typically, the elongation at break decreased, while the tensile strength increased with the increasing of RDX contents.

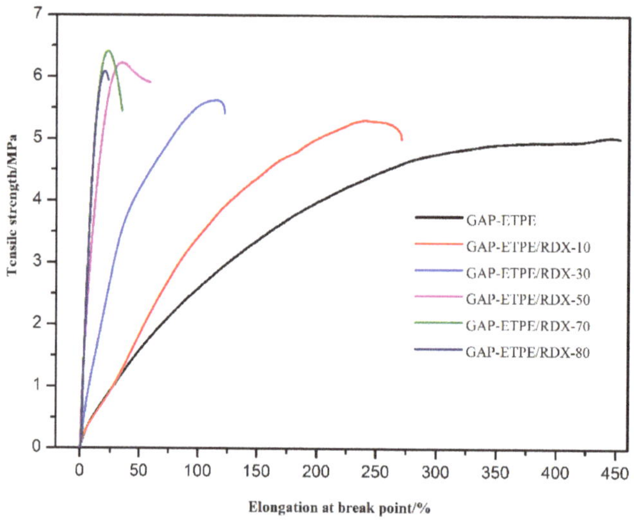

Figure 5. The tensile stress–strain curve of GAP–ETPE/RDX model propellants.

Table 2. Tensile strength, Young's Modulus, and tensile strain of the samples.

Sample Name	Tensile Strength/MPa	E/MPa	ε_b/%
GAP–ETPE	5.01 ± 0.52	1.1 ± 0.1	452.3 ± 11.2
GAP–ETPE/RDX-10	5.32 ± 0.73	2.2 ± 0.2	269.2 ± 8.3
GAP–ETPE/RDX-30	5.61 ± 0.51	4.6 ± 0.6	121.5 ± 14.1
GAP–ETPE/RDX-50	6.12 ± 0.32	7.6 ± 0.4	85.9 ± 3.5
GAP–ETPE/RDX-70	6.43 ± 0.65	20.1 ± 0.5	34.7 ± 4.6
GAP–ETPE/RDX-80	6.07 ± 0.58	29.5 ± 2.5	22.5 ± 2.1

Figure 6. SEM of tensile fracture surface of GAP–ETPE/RDX model propellants with various ratios, 50%, 70% and 80%, corresponding with (**a**–**c**).

3.3. Interaction-Enhanced Mechanism Analysis

In order to understand the reasons for the improved mechanical properties, the molecular structure of the GAP–ETPE and RDX was analyzed, and the electronic interaction model between them was constructed. As shown in Figure 7, the electronegativity of oxygen atom (O) in $-NO_2$ group of RDX is well reported to be stronger than the nitrogen atom (N) in the –CN group; O atom exhibits electrophilic property, N atom shows electron donor property, hence, the "induce effect" is generated between –CN and $-NO_2$ [5]. When RDX particles were filled into the GAP–ETPE binder containing the –CN group, the physical cross-linking structure was formed, and it produced stronger adhesion force between the GAP–ETPE chains and RDX particles [8]. Furthermore, the filler RDX is a six-membered structure with rigid features, which acted as reinforcement and caused the interactions to restrict the mobility of GAP–ETPE polymer chains with RDX percentages increasing.

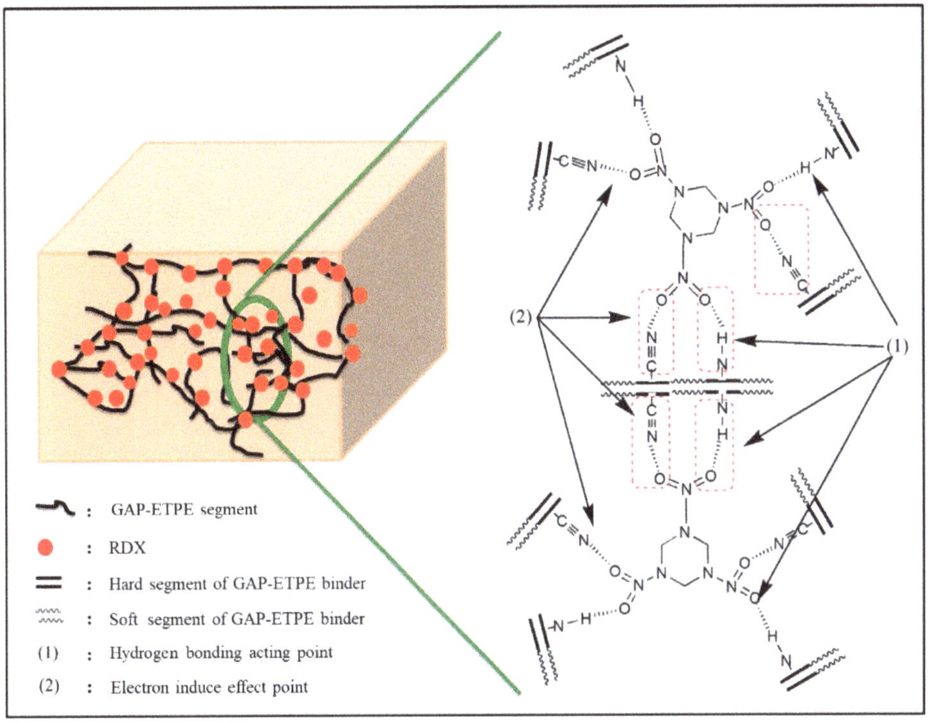

Figure 7. The induced-effect model between –CN and –NO_2.

3.4. *Thermal Decomposition Mechanism of GAP–ETPE/RDX*

The thermal decomposition of the GAP–ETPE/RDX model propellants is an important factor to predict safety and burning performance [23]. Therefore, it is necessary to understand the thermal decomposition process of pure GAP–ETPE and GAP–ETPE/RDX-50 model propellants.

(1) TG/DTG analysis of GAP–ETPE/RDX

Figure 8a,b show the TG results and DTG curves of GAP–ETPE, RDX, and GAP–ETPE/RDX, respectively. In the TGA curve of GAP–ETPE, two characteristic weight loss steps were observed. The first step (a) started at 233 °C, ended at 284 °C, and the maximum mass-losing peak temperature was located at 252 °C, which mainly corresponded to the decomposition of the -N_3 group. The second stage involved the decomposition of the polyether skeleton of the GAP–ETPE binder between 322 and 478 °C. As for RDX, there was only one loss step, which started at 192 °C and ended at 247 °C, and the maximum mass-loss peak temperature was about 240.8 °C. However, as for GAP–ETPE/RDX, there were four main stages of decomposition, which are clearly depicted in the DTG curve in the Figure 8b inset. We speculated that the first two stages are attributed to RDX decomposition, while the third and fourth are corresponding to the –N_3 groups and polyether skeleton of the GAP–ETPE binder, respectively.

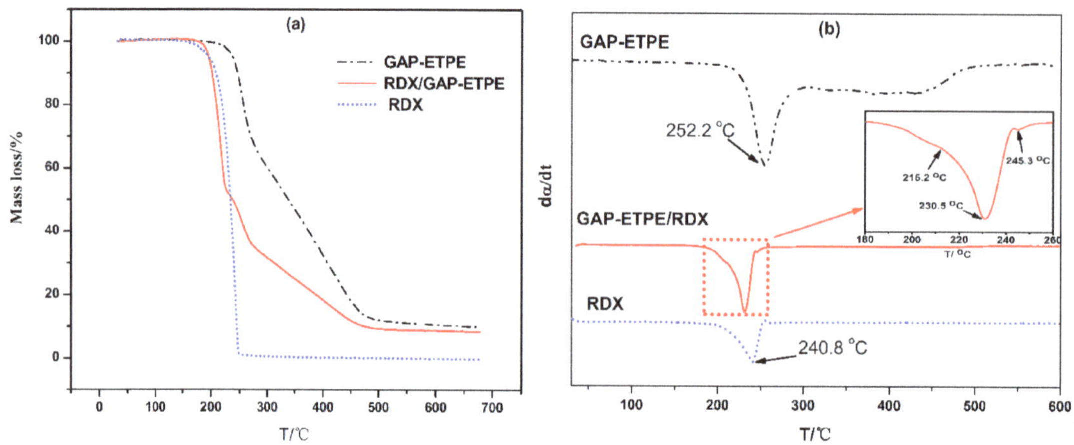

Figure 8. (**a**) TGA and (**b**) DTG curves of samples GAP–ETPE, GAP–ETPE/RDX, and RDX.

It can also be seen that the GAP–ETPE/RDX composites have lower degradation temperatures than neat RDX. This decrease is related to the reduced heat loss due to covering RDX by the binder [24]. The mixture of the GAP–ETPE binder presented a small decomposition peak located at about 213.5 °C, and it seemed to present a new phenomenon that the thermal decomposition stage of GAP–ETPE/RDX was divided into two stages between 200 and 240 °C, shown in Figure 8 inset, which have been never reported in the literature. We would like to clarify this phenomenon in the next step.

As expected, the thermal decomposition peak of $-N_3$ of GAP–ETPE/RDX is a lower temperature (246 °C) than neat GAP–ETPE. This may be caused by the heat generated from the decomposition of RDX, which acted as a heat source and accordingly caused it to preferentially absorb the heat of GAP–ETPE binder.

To understand the thermal decomposition mechanism, the overlapping peaks of the DTG curve were resolved into four stages by Gaussian fitting (Figure 9). By comparing the DTG of GAP–ETPE/RDX, we found that the degradation stage of GAP–ETPE/RDX between 170 and 240 °C corresponded to the decomposition of RDX, and the stage between 200 and 319 °C corresponded to the decomposition of the $-N_3$ of the GAP–ETPE binder. The last decomposition stage was related to the polyether and polyurethane skeleton of the GAP–ETPE binder. The peaks of the four stages were at 213.5, 228.1, 245.3, and 386.9 °C, respectively.

(2) TG/FTIR analysis of GAP–ETPE/RDX

Figure 10 shows the FTIR spectrum of gas products at respective decomposition peaks obtained from Figure 9. In Figure 10, the spectra at 215.2 °C shows the appearance of bands at about 2369, 2238, 2174, 1745, and 713 cm^{-1}, which corresponded to the characteristic absorption of CO_2, N_2O, CO, CH_2O, and HCN, respectively; these typical features mainly are attributed to the degradation of the RDX [25]. Thus, we can speculate that the thermal decomposition stage of RDX in GAP–ETPE/RDX is divided into two decomposition stages, which correspond to the slow rate decomposition and self-catalyzed accelerated decomposition stage, respectively. This phenomenon was verified by peak intensity from the TG/FTIR curve in Figure 10.

Furthermore, the gas products containing $-CH_2$ (2926 cm^{-1}), CO_2 (2380 cm^{-1}), N_2O (2237 cm^{-1}), $-N_3/HN_3$ (2100 cm^{-1}), and NH_3 (964 cm^{-1}) release at the temperature of maximum signal (about 245.3 °C) [26]. This is attributed to the degradation of the azide groups, corresponding to the third decomposition stage of the TG test. In addition, at 433.4 °C, the emission of CO_2, $-CH_2$, and HCHO are detected by the appearance of absorption bands at about 2300, 2926, and 1112 cm^{-1}, respectively, indicating that the main chain of the GAP–ETPE binder decomposed at this temperature [27]. When the temperature reached 550 °C, all characteristic peaks disappeared, indicating that GAP–ETPE/RDX decomposed completely.

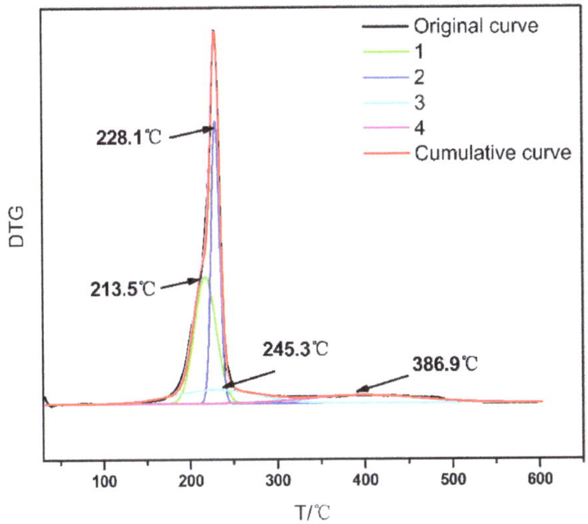

Figure 9. Fitting curve of DTG of GAP–ETPE/RDX.

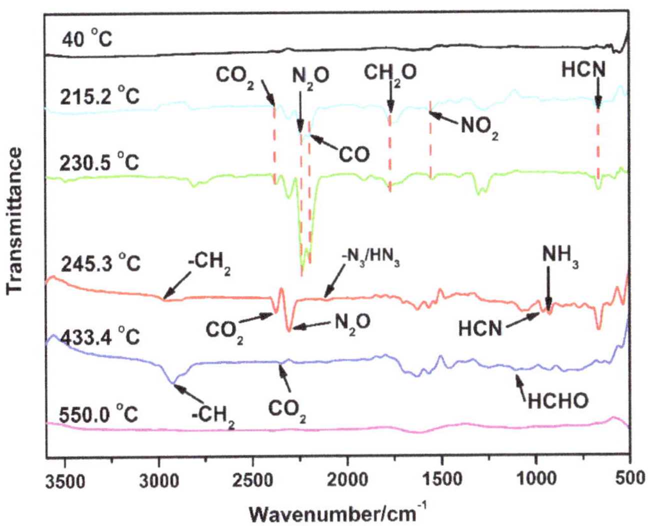

Figure 10. FTIR spectra of gas products during decomposition at respective peaks.

Based on the above analysis, we inferred the decomposition mechanism of GAP–ETPE/RDX, as shown in Figure 11. First of all, at 175–231 °C, the filler RDX thermally decomposed; hydrogen abstraction from RDX occurs through NO_2, leading to the formation of cis-HONO and a ring intermediate RDX_{-H} with carbon as radical center and subsequently decomposes to methylene nitramine CH_2NNO_2, HCN, and NO_2 [26]. Among of gas products, NO_2 possessed catalyst effect for further decomposition of RDX [26]. Hence, the process of RDX thermal decomposition was divided into two stages (slow rate decomposition and self-catalyzed accelerated decomposition stage) and released CO_2, N_2O,

CO, CH_2O, and HCN from GAP–ETPE/RDX. Secondly, at 240–310 °C, the heat from the RDX decomposition stage transferred into the GAP–ETPE binder, which prompted the $-N_3$ group decomposition and caused the maximum decomposition peak temperature ahead time. In this stage, $-CH_2$, CO_2, N_2O, $-N_3/HN_3$, and NH_3 was released because of the $-N_3$ group decomposition. Finally, at 320–500 °C, the polyether and polyurethane were further decomposed into low molecular weight ether and ketones, and at the same time, N_2O, HCN, NH_3, and CO_2 were released.

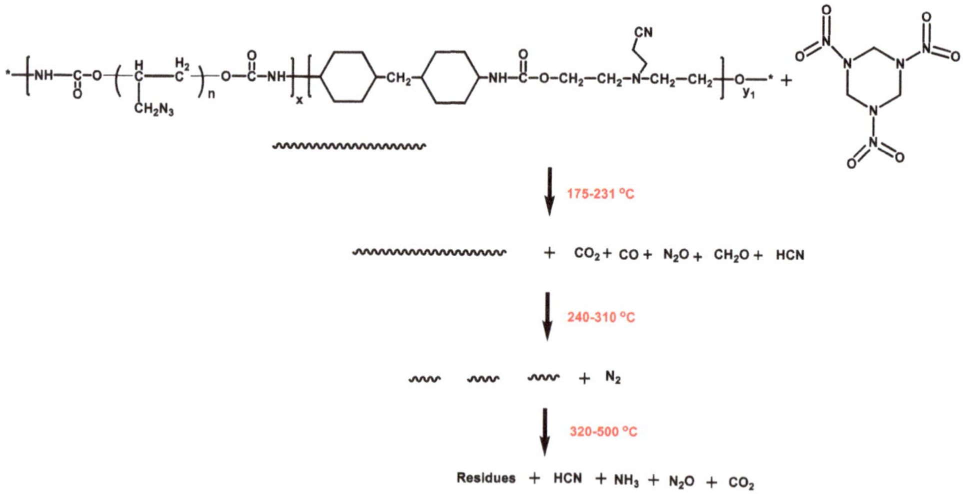

Figure 11. The thermal decomposition process of GAP–ETPE/RDX.

4. Conclusions

A novel GAP–ETPE binder with bonding functional effect was synthesized, and its composite with RDX was prepared. The mechanical and thermal decomposition performance of GAP–ETPE/RDX composites was investigated, and the following conclusions were obtained:

1. DMA data showed the E′ value of GAP–ETPE/RDX composite increased from 7280 MPa to 17,000 MPa, and the *Tg* value was shifted to higher temperatures, ranging from −30.1 °C to −26.1 °C with the RDX percentages.
2. The static mechanical test showed that GAP–ETPE/RDX composite did not display a de-wetting phenomenon because of the "induce effect" between –CN and $-NO_2$ group, which led to a stronger interfacial adhesion force for the GAP–ETPE/RDX model propellant.
3. The TG/FTIR results show that there are four stages during the GAP–ETPE/RDX thermal decomposition: the first two stages are attributed to RDX decomposition process (175–231 °C), which is divided into two stages, one is slow rate decomposition, and the other is self-catalyzed accelerated decomposition stage. The third stage is the azide group (240–310 °C), and the last stage corresponds to the polyether and polyurethane decomposition (320–500 °C).

Author Contributions: Methodology, Q.S. and X.-M.Y.; validation, Q.S.; formal analysis, Q.S.; investigation, Q.S., X.-M.Y. and G.-Z.Y.; data curation, Q.S.; writing—original draft, Q.S., X.-M.Y. and G.-Z.Y.; writing—review and editing, X.-M.Y. and G.-Z.Y.; supervision, X.-M.Y.; funding acquisition, G.-Z.Y. All authors have read and agreed to the published version of the manuscript.

Funding: This work was partially supported by NEWSAFE (No.: PID2022-143324NA-I00) Projects funded by Ministerio De Ciencia E Innovación (MINECO, Spain). This work was also funded in part by the University of Design, Innovation, and Technology (UDIT), code INC-UDIT-2024-APC07.

Institutional Review Board Statement: Not applicable.

Data Availability Statement: The original contributions presented in the study are included in the article; further inquiries can be directed to the corresponding authors.

Acknowledgments: Not applicable.

Conflicts of Interest: The authors declare no potential conflicts of interest with respect to the research, authorship, and/or publication of this article.

References

1. Sikder, A.K.; Reddy, S. Review on Energetic Thermoplastic elastomers (ETPEs) for military science. *Propellants Explos. Pyrotech.* **2013**, *38*, 14–28. [CrossRef]
2. Wan, M.; Shi, C.; Qian, X.; Qin, Y.; Jing, J.; Che, H.; Ren, F.; Li, J.; Yu, B.; Zhou, K. Design of novel double-layer coated ammonium polyphosphate and its application in flame retardant thermoplastic polyurethanes. *Chem. Eng. J.* **2023**, *459*, 141448. [CrossRef]
3. Miller, R.S. Research on new energetic materials. *J. Mater. Res. Soc. Symp. Proc.* **1996**, *418*, 3–14. [CrossRef]
4. Murphy, E.A.; Ntozakhe, T.; Murphy, C.J.; Fay, J.J.; Sperling, L.H.; Manser, G.E. Characterization of poly(3,3-bisethoxymethyl oxetane) and poly(3,3-bisazidomethyl oxetane) and their block copolymers. *J. Appl. Polym. Sci.* **1989**, *37*, 267–281. [CrossRef]
5. Zhang, Z.; Luo, N.; Deng, J.; Ge, Z.; Luo, Y. A kind of bonding functional energetic thermoplastic elastomers based on glycidyl azide polymer. *J. Elastomers Plast.* **2016**, *48*, 728–738. [CrossRef]
6. Guo, M.; Ma, Z.; He, L.; He, W.; Wang, Y. Effect of varied proportion of GAP-ETPE/NC as binder on thermal decomposition behaviors, stability and mechanical properties of nitramine propellants. *J. Therm. Anal. Calorim.* **2017**, *130*, 909–918. [CrossRef]
7. Min, B.S.; Ko, S.W. Characterization of segmented block copolyurethane network based on glycidyl azide polymer and polycaprolactone. *Macromole Res.* **2007**, *15*, 225–233. [CrossRef]
8. Sun, Q.; Sang, C.; Wang, Z.; Luo, Y. The study of mechanical and creep properties of glycidyl azide polyol energetic thermoplastic elastomer binder with bonding group with RDX and its interface reinforcement mechanism. *Mater. Res. Express* **2018**, *5*, 025309. [CrossRef]
9. Wang, Z.; Zhang, T.; Zhang, Z.; Ge, Z.; Luo, Y. Effect of hard-segment content on rheological properties of glycidyl azide polyol-based energetic thermoplastic polyurethane elastomers. *Polym. Bull.* **2016**, *73*, 3095–3104. [CrossRef]
10. Wang, Z.; Zhang, T.; Zhao, B.; Luo, Y. Effect of nitrocellulose (NC) on morphology, rheological and mechanical properties of glycidyl azide polymer based energetic thermoplastic elastomer/NC blends. *Polym. Int.* **2017**, *66*, 705–711. [CrossRef]
11. Louafi, E.; Boulkadid, M.K.; Belgacemi, R.; Touidjine, S.; Akbi, H.; Belkhiri, S.; Benaliouche, F. Decomposition reaction kinetics of double-base propellant catalyzed with graphene oxide–copper oxide nanocomposite. *Int. J. Chem. Kinet.* **2023**, *55*, 479–488. [CrossRef]
12. Yang, W.; Liu, W.; Zheng, M.; Zhang, X.; Jin, P.; Li, T.; Luo, Y. Preparation of PDA@ AP composite particles by two-step method and its application in glycidyl azide polymer-energetic thermoplastic elastomer propellants. *J. Appl. Polym. Sci.* **2023**, *140*, e54654. [CrossRef]
13. Nardai, M.M. *Cohesion Properties in PBX and Composite Propellants Computational Results and Experimental Aspects*; Fraunhofer ICT: Pfinztal, Germany; Karlsruhe, Germany, 2015.
14. Ding, Y.; Hu, C.; Guo, X.; Che, Y.; Huang, J. Structure and mechanical properties of novel composites based on glycidyl azide polymer and propargyl-terminated polybutadiene as potential binder of solid propellant. *J. Appl. Polym. Sci.* **2014**, *131*, 40007. [CrossRef]
15. Kimura, E.; Oyumi, Y. Effects of copolymerization ratio of BAMO/NMMO and catalyst on sensitivity and burning rate of HMX propellant. *Propellants Explos. Pyrotech.* **1995**, *20*, 215–221. [CrossRef]
16. Zhang, Z.; Wang, G.; Wang, Z.; Zhang, Y.; Ge, Z.; Luo, Y. Synthesis and characterization of novel energetic thermoplastic elastomers based on glycidyl azide polymer (GAP) with bonding functions. *Polym. Bull.* **2015**, *72*, 1835–1847. [CrossRef]
17. Sun, Q.; Sang, C.; Wang, Z.; Luo, Y. Improvement of the creep resistance of glycidyl azide polyol energetic thermoplastic elastomer-based propellant by nitrocellulose filler and its mechanism. *J. Elastomers Plast.* **2018**, *50*, 579–595. [CrossRef]
18. *GB/T 528-1998*; Rubber, Vulcanized or Thermoplastic Determination of Tensile Stress-Strain Properties. Standards Press of China: Beijing, China, 1999.
19. Gerbase, A.E.; Petzhold, C.L.; Costa, A.O. Dynamic mechanical and thermal behavior of epoxy resins based on soybean oil. *J. Am. Oil Chem. Soc.* **2002**, *8*, 797–802. [CrossRef]
20. Li, M.; Cho, U.R. Effectiveness of coupling agents in the poly (methyl methacrylate)-modified starch/styrene-butadiene rubber interfaces. *Mater. Lett.* **2013**, *92*, 132–135. [CrossRef]
21. Pan, B.; Luo, Y.; Tan, H. Study on Adhesion Properties of CL-20 and Tree molecular Bonding Agent. *Energ. Mater.* **2004**, *12*, 199–202.
22. Khoury Moussa, H.; Challita, G.; Badreddine, H.; Montay, G.; Guelorget, B.; Vallon, T.; Yared, W.; Abi Rizk, M.; Alhussein, A. Enhancement of mechanical properties of high modulus polypropylene grade for multilayer sewage pipes applications. *J. Appl. Polym. Sci.* **2023**, *140*, e53314. [CrossRef]

23. Bauetdinov, Y.; Grekova, A.; Sangwan, R. Thermal stability and decomposition mechanisms of hexatetracarbon: Tight-binding molecular dynamics and density functional theory study. *Mod. Phys. Lett. B* **2023**, *37*, 2350023-1. [CrossRef]
24. Felix, S.P.; Singh, G.; Sikder, A.; Aggrawal, J. Studies on energetic compounds: Part 33: Thermolysis of keto-RDX and its plastic bonded explosives containing thermally stable polymers. *Thermochim. Acta* **2005**, *426*, 53–60. [CrossRef]
25. Patidar, L.; Thynell, S.T. Quantum mechanics investigation of initial reaction pathways and early ring-opening reactions in thermal decomposition of liquid-phase RDX. *Combust. Flame* **2017**, *178*, 7–20. [CrossRef]
26. Zhao, Y.; Luo, Y.; Li, X.; Li, G. Kinetics and mechanism of thermal decomposition reaction of BAMO/GAP tri-block copolymer. *Polym. Mater. Sci. Eng.* **2012**, *11*, 42–45.
27. Zhang, Z.; Wang, G.; Luo, N.; Huang, M.; Jin, M.; Luo, Y. Thermal decomposition of energetic thermoplastic elastomers of poly (glycidyl nitrate). *J. Appl. Polym. Sci.* **2014**, *131*. [CrossRef]

polymers

MDPI

Article

Material Performance Evaluation for Customized Orthoses: Compression, Flexural, and Tensile Tests Combined with Finite Element Analysis

Daniela Trindade [1,2,3,*], Rachel Habiba [1,4], Cristiana Fernandes [1], André A. Costa [1], Rui Silva [1,5], Nuno Alves [1,6], Rui Martins [7,8], Cândida Malça [1,9], Ricardo Branco [10] and Carla Moura [1,2,11,*]

1 Center for Rapid and Sustainable Product Development (CDRSP), Polytechnic of Leiria, 2430-028 Marinha Grande, Portugal; rachel.d.habiba@ipleiria.pt (R.H.); cristiana.h.fernandes@gmail.com (C.F.); rui.d.silva@ipleiria.pt (R.S.); nuno.alves@ipleiria.pt (N.A.); candida@isec.pt (C.M.)
2 Applied Research Institute, Polytechnic Institute of Coimbra, Rua da Misericórdia, Lagar dos Cortiços, S. Martinho do Bispo, 3045-093 Coimbra, Portugal
3 Abel Salazar Biomedical Sciences Institute (ICBAS), University of Porto (UP), Rua de Jorge Viterbo Ferreira, No. 228, 4050-313 Porto, Portugal
4 Department of Mechanical Engineering, University of Coimbra, Rua Luis Reis Santos, 3030-788 Coimbra, Portugal
5 CIPER, Faculdade de Motricidade Humana, Universidade de Lisboa, 1495 Cruz Quebrada Dafundo, 1649-004 Lisbon, Portugal
6 Associate Laboratory for Advanced Production and Intelligent Systems (ARISE), 4050-313 Porto, Portugal
7 UNIDEMI, Department of Mechanical and Industrial Engineering, Nova School of Science and Technology, Universidade NOVA de Lisboa, Campus de Caparica, 2829-516 Caparica, Portugal; rfspm@fct.unl.pt
8 Laboratório Associado de Sistemas Inteligentes (LASI), 4800-058 Guimarães, Portugal
9 Coimbra Institute of Engineering (ISEC), Polytechnic Institute of Coimbra, Rua Pedro Nunes, Quinta da Nora, 3030-199 Coimbra, Portugal
10 CEMMPRE-ARISE, Department of Mechanical Engineering, University of Coimbra, Rua Luis Reis Santos, 3030-788 Coimbra, Portugal; ricardo.branco@dem.uc.pt
11 Research Centre for Natural Resources Environment and Society (CERNAS), Polytechnic Institute of Coimbra, Bencanta, 3045-601 Coimbra, Portugal
* Correspondence: daniela.trindade@ipleiria.pt (D.T.); carla.moura@ipc.pt (C.M.)

Abstract: Orthoses are commonly used for treating injuries to improve the quality of life of patients, with customized orthoses offering significant benefits. Additive manufacturing, especially fused deposition modelling, enhances these benefits by providing faster, more precise, and more comfortable orthoses. The present study evaluates nine polymeric materials printed in horizontal and vertical directions by assessing their performance through compressive, flexural, and tensile tests. Among all materials, polycarbonate, polylactic acid, and ULTEM™ 1010 showed the most promising results, not only because they had the highest mechanical values, but also due to their minimal or no difference in performance between printing directions, making them advantageous in orthoses fabrication. Based on this, a finite element model of an ankle–foot orthosis was developed to simulate the deformation, strain, and stress fields under static conditions. The findings aim to optimize material selection for orthotic fabrication, where ULTEM™ 1010 is presented as the material with improved performance and durability.

Keywords: customized orthoses; additive manufacturing; polymeric materials; ankle–foot orthosis; mechanical properties; static conditions

1. Introduction

Orthopaedic devices are commonly used for treating injuries that can be caused by falls, age-related illnesses, or accidents. Orthoses are a type of assistive device that can be used in patients with physical impairments. The main function of these devices is to provide support and correct a certain segment of the body, confine joint movement, and minimize the risk of malformations by distributing the loading forces [1,2]. They can be categorized depending on (I) the body portion: upper limb, spinal, and lower limb, or (ii) the joint involved: wrist–hand, lumbar, and ankle–foot [3].

Customized orthoses present good outcomes in patients, such as comfort and pain reduction [4], but as they are handmade, their quality highly depends on the competence and expertise of the specialist [5]. Additive manufacturing (AM) presents several advantages to the time-consuming and laborious conventional fabrication of custom orthoses, such as plaster casting [3,6]. The production of the orthoses is faster; the patient's experience is more comfortable since scanners can be used to aid in obtaining the desired geometry; the number of technicians and the manual work is reduced; the model of the orthoses can be archived and reproduced when necessary; and there is less need for production equipment, therefore less storage space [7].

Fused deposition modelling (FDM) is an AM technology that allows the production of three-dimensional objects through the extrusion of a material layer by layer. This technique allows for high precision in creating complex geometries that are challenging to achieve with traditional methods. Consequently, it can enhance orthoses' performance, durability, and modern aesthetics [8–10]. Furthermore, FDM offers greater accuracy, ease of use, and cost-effectiveness compared to other AM strategies, such as selective laser sintering [11]. Despite these advantages, the benefits of FDM technology are still underexplored [3]. One major challenge is selecting the right material for orthoses, which must meet various mechanical and physical properties [5,12,13]. No single material can meet all adequate criteria, but the final product should be lightweight, user-friendly, cost-effective, durable, body-compatible, and suitable for its intended use (e.g., rehabilitation or support). Using a hard material or an improper design can result in an uncomfortable or biomechanically incorrect orthosis [1,5,13–15].

The primary objective of this study is to evaluate nine polymeric materials printed in two printing directions (horizontal and vertical relative to the base plate) by analysing their compressive, flexural, and tensile properties. The second objective is the development of a three-dimensional finite element model of a real ankle–foot orthosis for simulating its deformation, stress, and strain fields under static loading conditions considering daily usage.

2. Materials and Methods

2.1. Materials Production

The study evaluated nine polymeric materials: acrylonitrile butadiene styrene (ABS), Nylon 12, polycarbonate (PC), polycarbonate/acrylonitrile butadiene styrene (PC-ABS), polyethylene terephthalate glycol (PETG), polylactic acid (PLA), thermoplastic polyurethane (TPU), and high-performance polyetherimide (PEI) thermoplastics ULTEM™ 1010, and ULTEM™ 9085. Materials were provided by Stratasys and the specimens were manufactured using a 3D printer by FDM (Stratasys F170 printer, Stratasys, Eden Prairie, MN, USA) with an infill density of 100%, infill angle of 45°, and a slice height of 254 µm. Each material was printed in two directions: horizontal (H) and vertical (V) relative to the base plate, as shown in Figure 1.

2.2. Tensile Testing

Tensile tests were conducted following ASTM D638-14 standards [16]. The tested specimens were printed in both H and V orientations. The tests were performed using a universal testing machine (Instron Model 5544, Norwood, MA, USA) equipped with a 100 kN load cell. The test speed was set to 5 mm/min.

2.3. Flexural Testing

Flexural tests were conducted using the same universal testing machine, but with a speed test of 2 mm/min according to ISO 178 standard [17]. The specimens used for this test were the same as those used in the tensile test since their specifications correspond to those used for this standard.

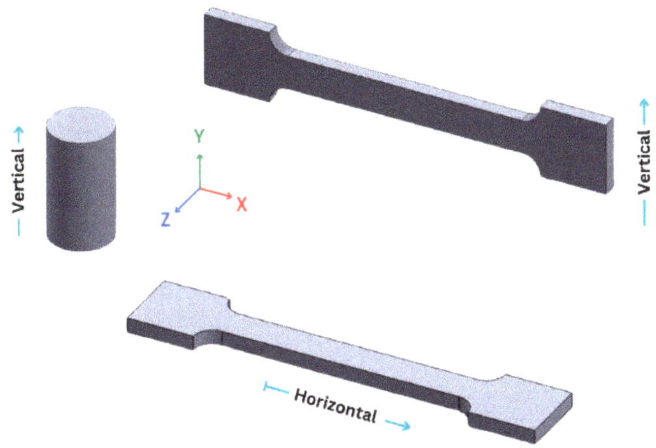

Figure 1. Horizontal (H) and vertical (V) printing directions of the tested specimens.

2.4. Compression Testing

Compression tests were performed according to ASTM D695-23 standards [18]. The cylindrical specimens tested were printed in the V direction. Testing was conducted using the same universal testing machine, with a speed test of 1 mm/min.

2.5. Data Analysis of the Mechanical Assays

Tests were conducted at room temperature and for each test type and material, five specimens were tested to ensure statistical reliability. The tensile/flexural/compressive strength, Young's modulus, and strain at break were recorded for each specimen; the results were averaged, and the standard deviations were calculated. The influence of printing direction on flexural and tensile properties was evaluated on GraphPad Prism 9 software with multiple unpaired t-tests. All tests were calculated with a confidence interval of 95%, where statistically significant differences are represented by * $p < 0.05$, ** $p < 0.01$ and *** $p < 0.001$. Correlations for the mechanical assays were also calculated with a Pearson correlation test on GraphPad using the same software and confidence interval above-mentioned.

2.6. Static Structural Test

The static structural analysis of the ankle–foot orthosis was performed for PC, PLA, and ULTEM™ 1010 due to their minimal or low differences in printing direction mechanical results. In this analysis, a real ankle–foot orthosis was simulated. The three-dimensional model was created using SolidWorks 2023, a software from Dassault Systèmes Corporation (Waltham, MA, USA). The model was imported as a Parasolid file (.x_t) into Ansys Workbench 19.2 software (Canonsburg, PA, USA), which provides a common platform integrating various Ansys applications for multi-physics simulations and design optimization. The finite element mesh contained 23,440 nodes and 11,758 elements, the element size was set at 5 mm, and the mesh type was tetrahedral. The physical model and the corresponding assembled meshed can be seen in Figure 2, where different perspective views of the ankle–foot orthosis are shown. The ankle–foot orthosis was designed with an increase in the length of the lever arm and the calf surface area to assure comfort and efficiency [19].

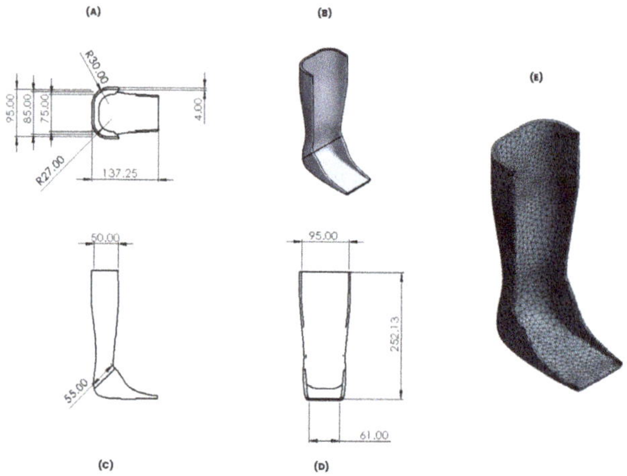

Figure 2. Ankle–foot orthosis design model: Top view (**A**), 3D projection view (**B**), front view (**C**), right view (**D**), and mesh model (**E**).

The simulation of the real-life effects can be seen in Figure 3: the area where the foot will be placed was assigned a ground-to-part relation with a fixed joint (in blue); to simulate the contact and force that the body may apply on the ankle–foot orthosis in real life when subjected to static conditions, a force of 490.03 N was used with force vector components (−3, 5, 490) N in X, Y, and Z directions. The force was applied to the entire model as shown in Figure 3. The applied force is according to the investigations of Marques et al. [20] and Ali et al. [21] describing the full contact moment in the gait cycle when the sole fully touches the ground.

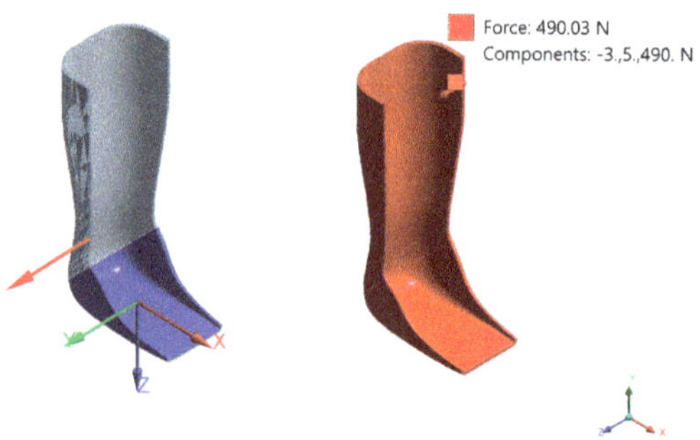

Figure 3. Fixed joint ground to part in blue and applied force in red.

The finite element model was assumed to be linear-elastic, homogeneous, and isotropic. The information about the isotropic elasticity, yield, and ultimate strength of the tested materials according to the material's supplier (Stratasys, Eden Prairie, MN, USA) is shown in Table 1.

Table 1. Material data.

Material	Young's Modulus (MPa)	Poisson's Ratio	Tensile Yield Strength (MPa)	Tensile Ultimate Strength (MPa)
PC	2250	0.39	57.9	57.3
PLA	3039	0.39	45.0	48.0
ULTEM™ 1010	2770	0.36	64.0	81.0

A static structural analysis was performed to obtain results relative to the total deformation, equivalent elastic strain, equivalent von Mises stress, and factor of safety defined based on maximum equivalent stress theory and tensile yield. The structural analyses were carried out for the three materials used in the numerical simulations.

3. Results

3.1. Tensile Tests

The tensile properties of the nine polymeric materials, printed in both H and V directions were evaluated. The three parameters analysed were tensile strength (Figure 4A), tensile Young's modulus (Figure 4B), and strain at break (Figure 4C). For tensile strength in the H direction, ULTEM™ 1010 gave the highest value of 69.99 ± 1.23 MPa, and TPU was the lowest with a value of 3.97 ± 0.03 MPa. For the V direction, ULTEM™ 8095 gave the highest value with 73.17 ± 0.33 MPa and TPU led to the lowest with a value of 4.36 ± 0.03 MPa. Statistically significant differences were found for most of the materials when comparing the printing directions, such as ABS, PC-ABS, PETG, Nylon12, TPU, and ULTEM™ 9085, where the V direction was the one with the highest values.

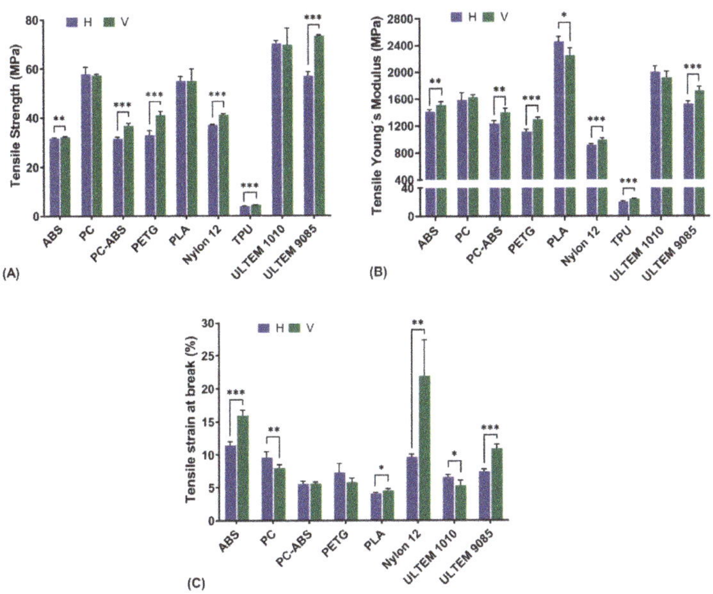

Figure 4. Tensile strength (**A**), Young's modulus (**B**), and strain at break (**C**) for all materials in both printing directions, horizontal (H) and vertical (V). Statistical analysis was conducted with multiple unpaired t-tests, and differences are represented by * $p < 0.05$, ** $p < 0.01$, and *** $p < 0.001$.

For tensile Young's modulus, TPU was the material with the lowest value for both directions, with a modulus of 20.04 ± 0.93 MPa and 24.18 ± 0.54 MPa for the H and V directions, respectively. The highest values were found in PLA, with a tensile Young's modulus of 2451.36 ± 81.12 MPa and 2245.74 ± 114.80 MPa, for the H and V direction, respectively. Statistically significant differences were found between directions for the same materials as for tensile strength, where the V direction was the one with the highest values, except for PLA where the H direction gave rise to a higher modulus.

Finally, for strain at break, TPU was the only material that did not lead to a break fracture. PLA was the material with the lower extension with a value of 4.07 ± 0.17% and 4.48 ± 0.31% for the H and V direction, respectively. For the higher values, in the H direction, ABS presented an extension of 11.41 ± 0.60%, and in the V direction, Nylon presented an extension of 21.89 ± 5.46%. Comparing printing directions, all materials led to statistically significant differences, except PC-ABS and PETG. For ABS, PLA, Nylon, and ULTEM™ 9085, the V direction led to higher values, whereas for PC and ULTEM™ 1010, the maximum values were found in the H direction.

3.2. Flexural Tests

Similar to tensile testing, the flexural strength (Figure 5A), flexural Young's modulus (Figure 5B), and flexural strain at break (Figure 5C) were evaluated for the nine tested materials in both H and V printing directions. It should be noted that in this research it was not possible to evaluate the TPU in our equipment due to its high flexibility, which led to some instability issues resulting in very unreliable graphs.

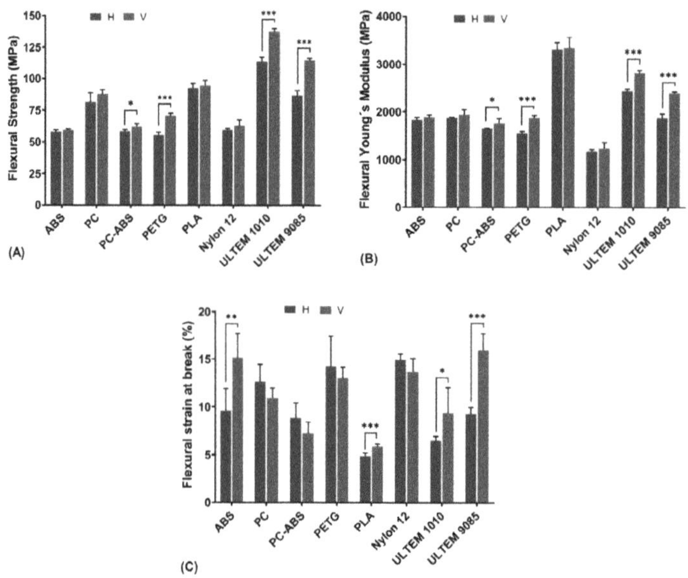

Figure 5. Flexural strength (**A**), Young´s modulus (**B**), and strain at break (**C**) for all materials in both printing directions, horizontal (H) and vertical (V). Statistical analysis was conducted with multiple unpaired t-tests, and differences are represented by * $p < 0.05$, ** $p < 0.01$, and *** $p < 0.001$.

For the flexural strength in the H direction, the highest value was attributed to ULTEM™ 1010 with a strength of 114 ± 3.27 MPa, and the lowest value to PETG with a value of 55.56 ± 2.26 MPa. In the V direction, ULTEM™ 9085 led to the highest value of 115 ± 1.44 MPa, and ABS exhibited the lowest value with 59.18 ± 0.92 MPa. Focusing on printing direction, statistically significant differences were found for the PC-ABS, PETG, ULTEM™ 1010, and ULTEM™ 9085, where the V direction gave origin to higher values.

For flexural Young's modulus, in both printing directions, PLA gave the highest values, whereas Nylon gave the lowest values: PLA-H was 1181.00 ± 39.36 MPa, PLA-V was 1236.32 ± 127.28 MPa, Nylon-H was 3313.68 ± 142.03 MPa, and Nylon-V was 3343.27 ± 219.58 MPa. Between directions, the statistically significant differences were similar to flexural strength, where PC-ABS, PETG, ULTEM™ 1010, and ULTEM™ 9085 gave rise to higher values in the V direction.

Strain at break was lower for PLA, in both printing directions, with values of 4.86 ± 0.33% for the H direction, and 5.89 ± 0.26% for the V direction. Nylon led to an extension of 14.93 ± 0.36%, being the material with the higher value in the H direction, whereas the in the V direction was ULTEM™ 9085 with a value of 15.98 ± 1.77%. Once again, TPU also did not lead to a fracture. Between printing directions, statistically significant differences were found for ABS, PLA, ULTEM™ 1010, and ULTEM™ 9085, where the V direction was the one with the higher values.

3.3. Compressive Tests

Compression properties were only evaluated in the V direction. The values of compressive strength, compressive Young's modulus, and compressive strain at break obtained in the tests are displayed in Figures 6A, 6B and 6C, respectively.

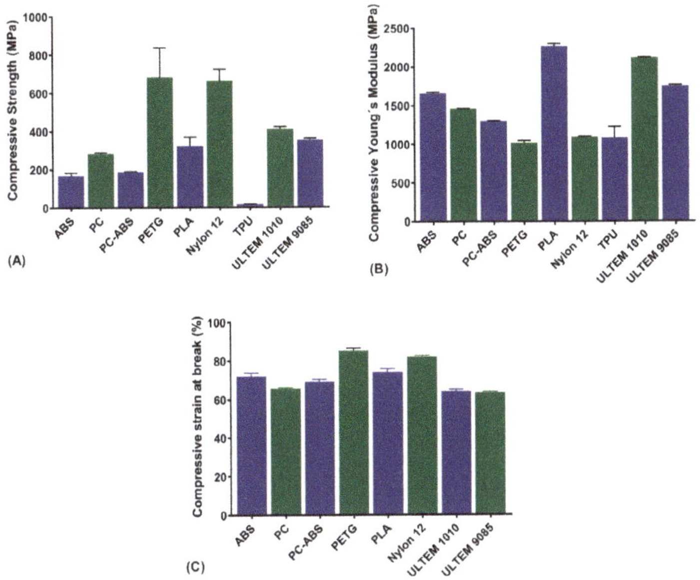

Figure 6. Compression strength (**A**), Young´s modulus (**B**), and strain at break (**C**) for all materials.

As far as the compression strength is concerned, the material that led to higher values was PETG with 680.7 ± 155.1 MPa and the lower was TPU with 16.9 ± 1.1 MPa. For compression Young's modulus, PLA had the higher values with 2264.0 ± 34.0 MPa, and PETG exhibited the lower value with 1008.8 ± 38.3 MPa. For compressive strain at break, the material that had a higher strain value was PETG with 85.1 ± 1.3%, while ULTEM™ led to the lower compressive strain at break with 63.2 ± 0.7 MPa.

3.4. Correlation Assays

Analysing the correlation studies for the tensile tests (see Figure 7(Ai,Aii)), it is possible to conclude that the strain at break is not associated with the tensile strength for both printing directions as the correlation coefficients are close to 0 (r = −0.2900 and r = −0.2946, for H and V direction, respectively). On the contrary, the tensile Young´s modulus, showed a negative correlation with the tensile strain at break for both printing directions (r = −0.529 and r = −0.562 for H and V direction, respectively), meaning that when the tensile Young´s modulus increases, the strain at break decreases. This was also confirmed by the $p < 0.001$ which confirmed that this negative correlation is not due to random sampling. As for tensile strength versus tensile Young´s modulus, a positive correlation was found (r = 0.602, and r = 0.597 for H and V direction, respectively, and $p < 0.001$), meaning that when one parameter increases the other also increases.

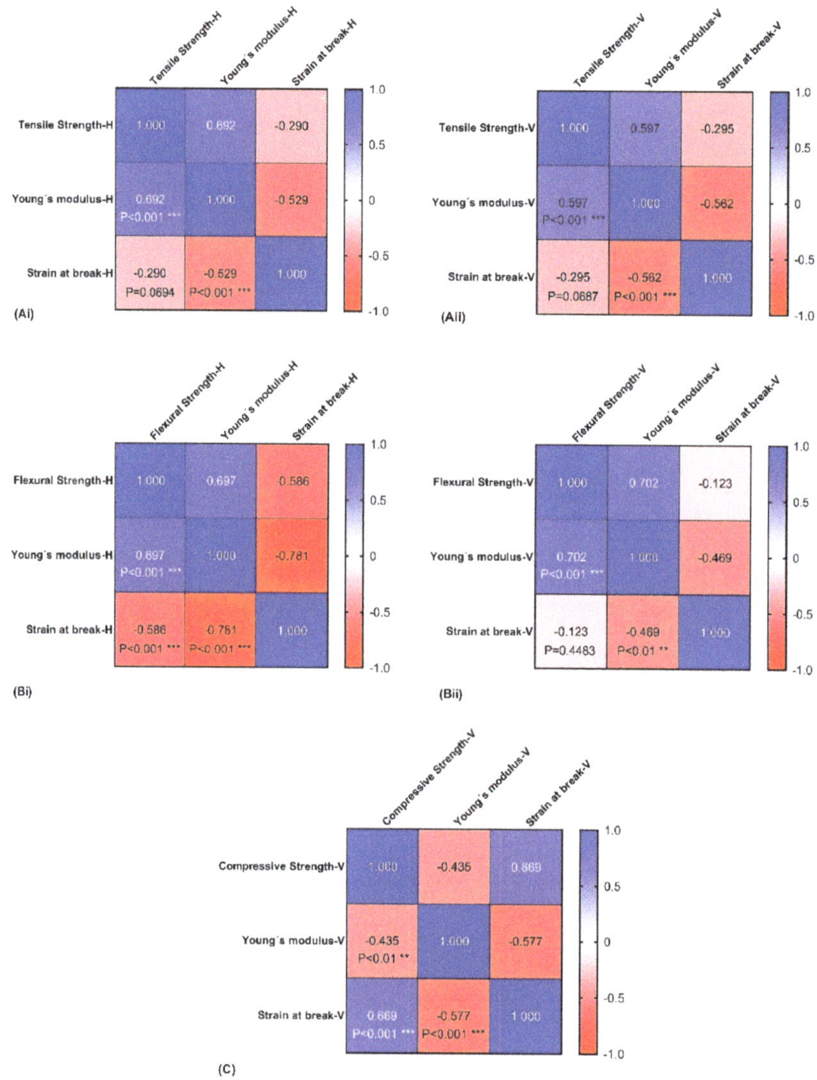

Figure 7. Correlation matrix for each mechanical assay: tensile test in the vertical (**Ai**) and horizontal direction (**Aii**), flexural test in the vertical (**Bi**) and horizontal direction (**Bii**), and compression test (**C**). The correlation coefficient is presented, as well as statistical differences by ** $p < 0.01$ and *** $p < 0.001$.

Regarding the flexural tests (see Figure 7(Bi,Bii)), similar results with the tensile assays were found for the V direction. Flexural strength versus flexural strain at break presented no relationship (r = −0.123, NS), flexural Young´s modulus versus flexural strain at break presented a negative correlation (r = −0.469, $p < 0.01$), and flexural strength versus flexural Young´s modulus presented a positive correlation (r = 0.702, $p < 0.001$). As for the H direction, interestingly, flexural strength versus flexural strain at break presented a negative correlation (r = −0.586, $p < 0.001$). The remaining analyses were similar to the V direction, as flexural Young´s modulus versus flexural strain at break presented a negative correlation (r = −0.781, $p < 0.001$), despite being a much stronger correlation with value close to

−1. Finally, flexural strength versus flexural Young´s modulus also presented a positive correlation (r = 0.697, $p < 0.001$).

For compression assays (see Figure 7C), similar to the other mechanical tests, compressive Young's modulus presented a negative association with compressive strain at break (r = −0.577, $p < 0.001$). The differences were found for the remaining correlations. Compressive strength versus compressive strain at break presented a positive correlation (r = 0.669, $p < 0.001$) and compressive strength versus compressive Young´s modulus presented a negative correlation (r = −0.435, $p < 0.01$).

3.5. Static Structural Analysis

Static structural analysis was carried out for PC, PLA, and ULTEM™ 1010 due to their minimal or low differences in mechanical results between printing directions, and also because they are the materials with the highest mechanical properties of all the materials, making them suitable for the production of orthotics. The results are represented in colour varying from blue to red, which correspond from the lower to the higher values of the plotted variable. For each material, equivalent stresses (Figure 8A), equivalent strains (Figure 8B), total deformation (Figure 8C), and safety factors (Figure 8D) are presented. Regarding stresses, for the three materials (PC, PLA, and ULTEM™ 1010), the maximum von Mises stresses were around 25 MPa and stress concentrations were more located in the area covering the ankle. The elastic strains were also more visible in that area, where PC showed the highest maximum elastic strain, followed by ULTEM™ 1010 and PLA. The upper area of the ankle–foot orthosis showed a significant deformation for the three materials, with red indicating maximum total deformation. PLA showed the lowest deformation compared to ULTEM™ 1010 and PC, while PC showed the highest deformation. All three materials demonstrated a minimum safety factor greater than 1. Among them, ULTEM™ 1010 had the highest safety factor, followed by PC. PLA showed the lowest safety factor. Table 2 summarises the main results obtained in the numerical simulations for the three materials (maximum von Mises stress, maximum elastic strain, maximum total deformation, and minimum safety factor).

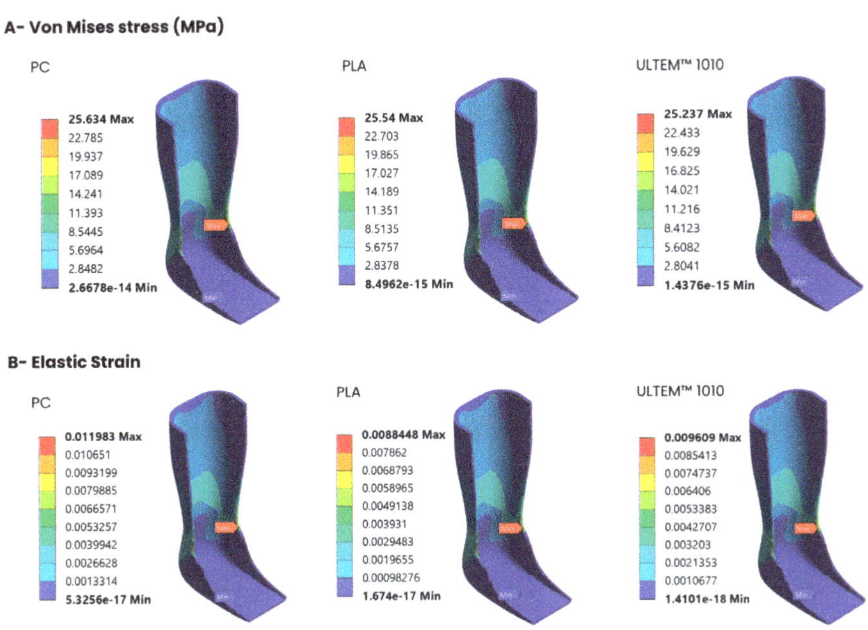

Figure 8. *Cont.*

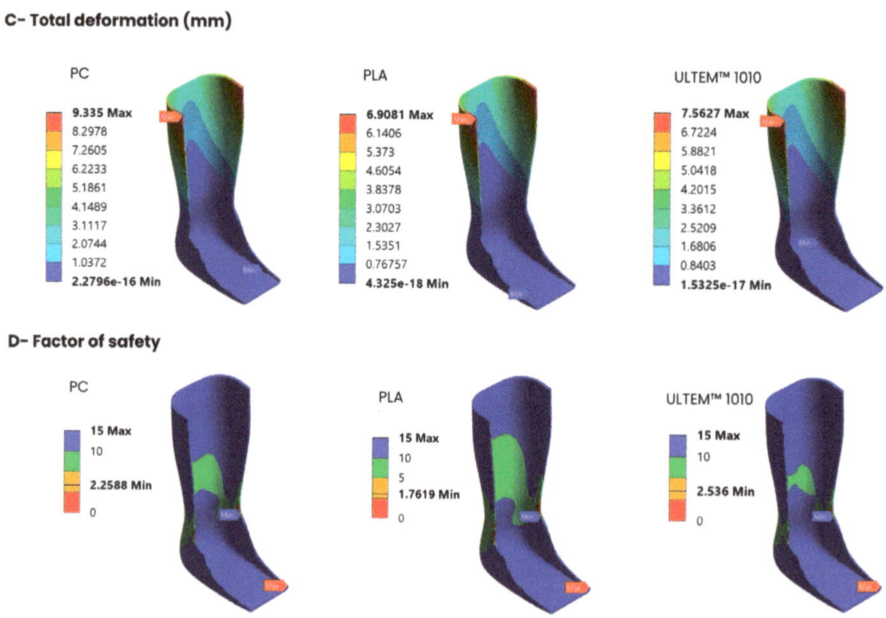

Figure 8. Static structural analysis of the ankle–foot orthosis: (**A**) Equivalent von Mises stress; (**B**) equivalent strain; (**C**) total deformation, and (**D**) safety factor for the three tested materials (PC, PLA, and ULTEM™ 1010).

Table 2. Result summary for the PC, PLA, and ULTEM™ 1010.

Material	Maximum Von Mises Stress (MPa)	Maximum Elastic Strain (mm/mm)	Maximum Total Deformation (mm)	Minimum Safety Factor
PC	25.63	11.98×10^{-3}	9.34	2.26
PLA	25.54	8.85×10^{-3}	6.91	1.76
ULTEM™ 1010	25.24	9.61×10^{-3}	7.56	2.54

4. Discussion

For orthotic AM production, it is essential that the chosen materials can withstand distinct mechanical stresses, including those resulting from flexural, compression, and tensile forces. These properties ensure that the orthosis will be durable and reliable for the patient while maintaining its structural integrity and functionality over time. Different authors have investigated different materials for orthotic production such as PC, PC-ABS, ULTEM, PLA, ABS, and PETG [22–29]. However, a consensus on the most suitable material is still debatable.

FDM-manufactured parts are known to be anisotropic due to the specificities inherent to this AM process, including the printing orientation [30,31]. This is why the mechanical properties of printed materials must be addressed in different orientations to achieve the desired results. Camargo et al. showed that the tensile and flexural strength of PLA-graphene material increases with the increase of the infill, while impact energy decreases. An increase in layer thickness also led to higher values in the referred mechanical properties [32]. Moreover, PLA also exhibited varying flexural strengths depending on the type of filling, such as rectangular, triangular, and honeycomb [33].

In the V direction, the layers of the printed materials are aligned parallel to the loads, while in the H direction, they are aligned perpendicularly. This characteristic resulted in better mechanical performance in the flexural tests, for all the analysed materials. For the tensile tests, the best performance was associated with the V direction, except for the PLA's Young´s modulus, and for PC and ULTEM™ 1010 strain at break, where the H direction showed higher values. Various studies have reported that printing directions affect the flexural properties of resins [34,35]. Similar findings have been reported for thermoplastics, aligning with the results found in the present study. The specimens printed parallel to the loads, presented higher flexural strength in ULTEM™ 9085 and ABS [36], and higher tensile strength in ABS [37]. The same was also observed for Nylon and ULTEM™ 9085 tensile strength, tensile Young´s modulus, and tensile strain at break [38,39]. Curiously, in the present study ULTEM™ 1010´s tensile strength and tensile Young´s modulus presented similar results between printing directions, but tensile strain at break was also higher for the H direction [39]. Although PLA [40] and PC [41] presented higher tensile strength values for specimens printed parallel to the loads in other studies, this was not observed in the present study. This discrepancy may occur likely due to variations in printing speed and temperature, which can affect the adhesion between layers and the consistency of the filament diameter and its quality, leading to differences in mechanical performance [6,42]. The rapid cooling from the FDM process can leave behind empty spaces due to a very rapid shrinkage of the material which leads to a deficiency in the adhesion between material layers, leading to residual stresses in the material [30].

The strength–ductility of the materials produced can be more effectively analysed through correlation studies of mechanical properties [43]. It is known that Young´s modulus is defined as the ability of a material to resist deformation [44]. The ultimate strength, used in this study as tensile/flexural/compression strength, is the maximum value that an object can resist without breaking [45], and strain at break is the point the material fractures [46]. Results of correlation demonstrated that there is no association between tensile strain at break and tensile strength. This means that the material´s ability to withstand stress in both printing directions does not predict its elongation. As for tensile Young´s modulus and tensile strain at break, there is a negative correlation. This means that materials with a higher Young´s modulus (stiffer materials) are often more brittle. Tensile strength versus tensile Young´s modulus presented a positive correlation, as both are related to the material´s ability to bear loads. For flexural tests, the same conclusions can be drawn, as similar results were obtained. The only difference was found in the H direction, where a negative correlation was found between flexural strength and flexural strain at break. In this direction, not only does a stiffer disc lead to a brittle material, but so does its load-bearing capacity. Lastly, the failure mechanisms of compressive loads led to differences when compared to the tensile and flexural tests: a positive correlation was found between compressive strength and compressive strain at break, meaning that the materials can withstand higher loads and also elongate more; and negative correlation between compressive strength and compressive Young´s modulus, where a material that can withstand more loads does not necessarily exhibit greater stiffness.

Regarding ankle–foot orthosis manufacturing, Raj et al. highlighted the advantages of using AM to produce ankle–foot orthosis compared to conventional manufacturing [47]. Overall PC, PLA, and ULTEM™ 1010 demonstrated the most promising outcomes. Not only do they present superior mechanical properties, but their consistent results in the different printing directions also make them particularly advantageous for orthosis fabrication due to a higher printing flexibility. Thus, they were selected for the simulations. The simulation of the ankle–foot orthosis designed in this study gives realistic results relative to its mechanical performance under real-world conditions while allowing for reduction of the amount of prototype iterations for validation. The results of static structural analysis give an insight into the mechanical performance of the ankle–foot orthoses produced using three distinct materials. The stress distribution patterns are comparable to each other allowing us to identify the concentrated high-stress regions where potential failures can occur. The maximum stresses for PC, PLA, and ULTEM™ 1010 are below their tensile yield strengths which ensures that the material behaves predictably, within its safe operating limits. PC showed the highest maximum elastic strain, which means that it is prone to deform more compared to ULTEM™ 1010 and PLA. The total deformation results show that PC is prone to significant deformation while PLA has a lower deformation, whereas ULTEM™ 1010 balances between them. Regarding the safety factor, a value lower than 1 indicates potential failure. In the three cases, the safety factor is higher than 1, which indicates that these materials are in the acceptable range. The safety factors of PC and ULTEM™ 1010 are greater than 2, indicating that the model can handle twice the force applied without failing. Based on the simulation results, it is clear that the current orthoses design will experience high stress levels in specific areas, regardless of the material used.

5. Conclusions

Through these tests, we can select materials that will optimize the performance of orthoses, contributing to better patient outcomes and satisfaction. The present study led to a better understanding of nine polymeric materials under various mechanical conditions. The correlation studies emphasized the importance of considering different mechanical properties for evaluating material performance. PC, PLA, and ULTEM™ 1010 presented the most interesting results because there were no differences in values between the print directions, making them more advantageous for orthosis printing. This led to the choice of a virtual ankle–foot orthosis based on these three materials. The FEA of the ankle–foot orthosis gives insight into the mechanical behaviour of an ankle–foot orthosis under static conditions. This result from the static structural analysis can help in optimizing ankle–foot orthoses for better performance under real loading conditions. Based on the numerical simulations, ULTEM™ 1010 exhibited the best performance.

Author Contributions: Conceptualization: D.T. and C.M. (Carla Moura); methodology; D.T., R.H., C.F., A.A.C., R.B. and C.M. (Carla Moura); formal analysis: D.T., R.H., C.F. and A.A.C.; validation: R.S., N.A., R.M., C.M. (Cândida Malça), R.B. and C.M. (Carla Moura); writing and revising of the article: D.T., R.H., C.F., A.A.C., R.S., N.A., R.M., C.M. (Cândida Malça), R.B. and C.M. (Carla Moura). All authors have read and agreed to the published version of the manuscript.

Funding: The authors acknowledge Fundação para a Ciência e a Tecnologia (FCT) for its financial support through the following projects: CDRSP-UIDB/04044/2020; CDRSP-UIDP/04044/2020; CEMMPRE-UIDB/00285/2020; UNIDEMI-UIDB/00667/2020; UNIDEMI-UIDP/00667/2020; and Associate Laboratory ARISE LA/P/0112/2020. This research was also funded through the institutional scientific employment program contract (CEEC-INST/00077/2021).

Institutional Review Board Statement: Not applicable.

Data Availability Statement: The original contributions presented in the study are included in the article; further inquiries can be directed to the corresponding author/s.

Conflicts of Interest: The authors declare no conflicts of interest.

References

1. Alqahtani, M.S.; Al-Tamimi, A.; Almeida, H.; Cooper, G.; Bartolo, P. A Review on the Use of Additive Manufacturing to Produce Lower Limb Orthoses. *Prog. Addit. Manuf.* **2020**, *5*, 85–94. [CrossRef]
2. Hensen, J.C.D.; Foggiatto, J.A.; Ulbricht, L.; Stadnik, A.M.W. Additive Manufacturing of Customized Lower Limb Orthoses—A Review. *Int. J. Innov. Educ. Res.* **2018**, *6*, 141–152. [CrossRef]
3. Wang, Y.; Tan, Q.; Pu, F.; Boone, D.; Zhang, M. A Review of the Application of Additive Manufacturing in Prosthetic and Orthotic Clinics from a Biomechanical Perspective. *Engineering* **2020**, *6*, 1258–1266. [CrossRef]
4. Aydin, L.; Kucuk, S. A Method for More Accurate FEA Results on a Medical Device Developed by 3D Technologies. *Polym. Adv. Technol.* **2018**, *29*, 2281–2286. [CrossRef]
5. Barrios-Muriel, J.; Romero-Sánchez, F.; Alonso-Sánchez, F.J.; Rodríguez Salgado, D. Advances in Orthotic and Prosthetic Manufacturing: A Technology Review. *Materials* **2020**, *13*, 295. [CrossRef] [PubMed]
6. Mian, S.H.; Abouel Nasr, E.; Moiduddin, K.; Saleh, M.; Alkhalefah, H. An Insight into the Characteristics of 3D Printed Polymer Materials for Orthoses Applications: Experimental Study. *Polymers* **2024**, *16*, 403. [CrossRef] [PubMed]
7. Pallari, J.H.P.; Dalgarno, K.W.; Munguia, J.; Muraru, L.; Peeraer, L.; Telfer, S.; Woodburn, J. Design and Additive Fabrication of Foot and Ankle-Foot Orthoses. In Proceedings of the 21st Annual International Solid Freeform Fabrication Symposium—An Additive Manufacturing Conference, Austin, TX, USA, 9–11 August 2010; pp. 9–11.

8. Chhikara, K.; Singh, G.; Gupta, S.; Chanda, A. Progress of Additive Manufacturing in Fabrication of Foot Orthoses for Diabetic Patients: A Review. *Ann. 3D Print. Med.* **2022**, *8*, 100085. [CrossRef]
9. Butt, J.; Bhaskar, R.; Mohaghegh, V. Investigating the Influence of Material Extrusion Rates and Line Widths on FFF-Printed Graphene-Enhanced PLA. *J. Manuf. Mater. Process.* **2022**, *6*, 57. [CrossRef]
10. Walbran, M.; Turner, K.; McDaid, A.J. Customized 3D Printed Ankle-Foot Orthosis with Adaptable Carbon Fibre Composite Spring Joint. *Cogent Eng.* **2016**, *3*, 1227022. [CrossRef]
11. Rengier, F.; Mehndiratta, A.; von Tengg-Kobligk, H.; Zechmann, C.M.; Unterhinninghofen, R.; Kauczor, H.-U.; Giesel, F.L. 3D Printing Based on Imaging Data: Review of Medical Applications. *Int. J. Comput. Assist. Radiol. Surg.* **2010**, *5*, 335–341. [CrossRef]
12. Wong, M.S.; Hassan Beygi, B.; Zheng, Y. Materials for Exoskeletal Orthotic and Prosthetic Systems. In *Encyclopedia of Biomedical Engineering*; Narayan, R., Ed.; Elsevier: Amsterdam, The Netherlands, 2019; pp. 352–367.
13. Sarma, T.; Pandey, D.; Sahai, N.; Tewari, R.P. Material Selection and Development of Ankle Foot Orthotic Device. *Mater. Today Proc.* **2019**, *18*, 2509–2514. [CrossRef]
14. Shahar, F.S.; Hameed Sultan, M.T.; Lee, S.H.; Jawaid, M.; Md Shah, A.U.; Safri, S.N.A.; Sivasankaran, P.N. A Review on the Orthotics and Prosthetics and the Potential of Kenaf Composites as Alternative Materials for Ankle-Foot Orthosis. *J. Mech. Behav. Biomed. Mater.* **2019**, *99*, 169–185. [CrossRef] [PubMed]
15. Portnova, A.A.; Mukherjee, G.; Peters, K.M.; Yamane, A.; Steele, K.M. Design of a 3D-Printed, Open-Source Wrist-Driven Orthosis for Individuals with Spinal Cord Injury. *PLoS ONE* **2018**, *13*, e0193106. [CrossRef]
16. *ASTM D638-14*; Standard Test Method for Tensile Properties of Plastics. ASTM International: West Conshohocken, PA, USA, 2022. [CrossRef]
17. *ISO 178:2019*; Plastics—Determination of Flexural Properties. ISO: Geneva, Switzerland, 2019.
18. *ASTM D695-23*; Standard Test Method for Compressive Properties of Rigid Plastics. ASTM International: West Conshohocken, PA, USA, 2023. [CrossRef]
19. Ferreira, P.D.P. Development of a Two-Dimensional Biomechanical Multibody Model for the Analysis of the Human Gait with an Ankle-Foot Orthosis/Desenvolvimento de Um Modelo Biomecânico Multibody Bidimensional Para a Análise Da Marcha Humana Com Uma Ortótese Do Tornozelo. Ph.D Thesis, Universidade do Minho, Braga, Portugal, 2012.
20. Marques, M.A.; Mendes, E.; Ramos, N.V.; Pinto, V.C.; Vaz, M.A. Finite-Element Analysis of Ankle-Foot Orthosis to Predict Fracture Conditions during Gait. In Proceedings of the 1st ICH Gaia, Porto, Portugal, January 2010; Available online: https://www.researchgate.net/publication/262684543_Finite_element_analysis_of_ankle_foot_orthosis_to_predict_fracture_conditions_during_gait (accessed on 30 July 2024).
21. Ali, M.H.; Smagulov, Z.; Otepbergenov, T. Finite Element Analysis of the CFRP-Based 3D Printed Ankle-Foot Orthosis. *Procedia Comput. Sci.* **2021**, *179*, 55–62. [CrossRef]
22. Blaya, F.; Pedro, P.S.; Silva, J.L.; D'Amato, R.; Heras, E.S.; Juanes, J.A. Design of an Orthopedic Product by Using Additive Manufacturing Technology: The Arm Splint. *J. Med. Syst.* **2018**, *42*, 54. [CrossRef]
23. Chen, R.K.; Chen, L.; Tai, B.L.; Wang, Y.; Shih, A.J.; Wensman, J. Additive Manufacturing of Personalized Ankle-Foot Orthosis. *Proc. NAMRI/SME* **2014**, *42*, 381–389.
24. Schrank, E.S.; Hitch, L.; Wallace, K.; Moore, R.; Stanhope, S.J. Assessment of a Virtual Functional Prototyping Process for the Rapid Manufacture of Passive-Dynamic Ankle-Foot Orthoses. *J. Biomech. Eng.* **2013**, *135*, 101011. [CrossRef]
25. Varga, P.; Lorinczy, D.; Toth, L.; Pentek, A.; Nyitrai, M.; Maroti, P. Novel PLA-$CaCO_3$ Composites in Additive Manufacturing of Upper Limb Casts and Orthotics—A Feasibility Study. *Mater. Res. Express* **2019**, *6*, 045317. [CrossRef]
26. Arulmozhi, R.S.; Vaidya, M.; Poojalakshmi, M.G.; Ashok Kumar, D.; Anuraag, K. 3D design and printing of custom-fit finger splint. *Biomed. Eng. Appl. Basis Commun.* **2018**, *30*, 1850032. [CrossRef]
27. Łukaszewski, K.; Wichniarek, R.; Górski, F. Determination of the Elasticity Modulus of Additively Manufactured Wrist Hand Orthoses. *Materials* **2020**, *13*, 4379. [CrossRef]
28. Santos, M.A.R.D.; Tokimatsu, R.C.; Treichel, T.L.E.; Prado, T.D.D.; CDS Junior, A. Orthosis and Prosthesis Development for Large and Medium Animals Using Reverse Engineering and Additive Manufacturing Techniques. *Int. J. Adv. Eng. Res. Sci.* **2020**, *7*, 169–177. [CrossRef]
29. Habiba, R.; Amaro, A.; Trindade, D.; Moura, C.; Silva, R.; Antão, A.; Martins, R.F.; Malça, C.; Branco, R. Comparative Analysis of Impact Strength among Various Polymeric Materials for Orthotic Production. *Polymers* **2024**, *16*, 1843. [CrossRef]
30. Butt, J.; Oxford, P.; Sadeghi-Esfahlani, S.; Ghorabian, M.; Shirvani, H. Hybrid Manufacturing and Mechanical Characterization of Cu/PLA Composites. *Arab. J. Sci. Eng.* **2020**, *45*, 9339–9356. [CrossRef]
31. Butt, J.; Bhaskar, R.; Mohaghegh, V. Non-Destructive and Destructive Testing to Analyse the Effects of Processing Parameters on the Tensile and Flexural Properties of FFF-Printed Graphene-Enhanced PLA. *J. Compos. Sci.* **2022**, *6*, 148. [CrossRef]
32. Camargo, J.C.; Machado, Á.R.; Almeida, E.C.; Silva, E.F.M.S. Mechanical Properties of PLA-Graphene Filament for FDM 3D Printing. *Int. J. Adv. Manuf. Technol.* **2019**, *103*, 2423–2443. [CrossRef]
33. Rebenaque, A.G.; González-Requena, I. Study of bending test of specimens obtained through fdm processes of additive manufacturing. *Procedia Manuf.* **2019**, *41*, 859–866. [CrossRef]
34. Derban, P.; Negrea, R.; Rominu, M.; Marsavina, L. Influence of the Printing Angle and Load Direction on Flexure Strength in 3D Printed Materials for Provisional Dental Restorations. *Materials* **2021**, *14*, 3376. [CrossRef] [PubMed]

35. Keßler, A.; Hickel, R.; Ilie, N. In Vitro Investigation of the Influence of Printing Direction on the Flexural Strength, Flexural Modulus and Fractographic Analysis of 3D-Printed Temporary Materials. *Dent. Mater. J.* **2021**, *40*, 641–649. [CrossRef] [PubMed]
36. Maroti, P.; Varga, P.; Abraham, H.; Falk, G.; Zsebe, T.; Meiszterics, Z.; Mano, S.; Csernatony, Z.; Rendeki, S.; Nyitrai, M. Printing Orientation Defines Anisotropic Mechanical Properties in Additive Manufacturing of Upper Limb Prosthetics. *Mater. Res. Express* **2018**, *6*, 035403. [CrossRef]
37. Dwiyati, S.T.; Kholil, A.; Riyadi, R.; Putra, S.E. Influence of Layer Thickness and 3D Printing Direction on Tensile Properties of ABS Material. *J. Phys. Conf. Ser.* **2019**, *1402*, 066014. [CrossRef]
38. Zaldivar, R.J.; Witkin, D.B.; McLouth, T.; Patel, D.N.; Schmitt, K.; Nokes, J.P. Influence of Processing and Orientation Print Effects on the Mechanical and Thermal Behavior of 3D-Printed ULTEM® 9085 Material. *Addit. Manuf.* **2017**, *13*, 71–80. [CrossRef]
39. Appalsamy, T.; Hamilton, S.L.; Kgaphola, M.J. Tensile Test Analysis of 3D Printed Specimens with Varying Print Orientation and Infill Density. *J. Compos. Sci.* **2024**, *8*, 121. [CrossRef]
40. Liu, H.; He, H.; Peng, X.; Huang, B.; Li, J. Three-dimensional Printing of Poly(Lactic Acid) Bio-based Composites with Sugarcane Bagasse Fiber: Effect of Printing Orientation on Tensile Performance. *Polym. Adv. Technol.* **2019**, *30*, 910–922. [CrossRef]
41. Akgümüş Gök, D. Determination of Optimum Printing Direction of the Polycarbonate Parts Produced by Additive Manufacturing. *Mater. Res. Express* **2024**, *11*, 045302. [CrossRef]
42. Tymrak, B.M.; Kreiger, M.; Pearce, J.M. Mechanical Properties of Components Fabricated with Open-Source 3-D Printers under Realistic Environmental Conditions. *Mater. Des.* **2014**, *58*, 242–246. [CrossRef]
43. Xie, X.; Bennett, J.; Saha, S.; Lu, Y.; Cao, J.; Liu, W.K.; Gan, Z. Mechanistic Data-Driven Prediction of as-Built Mechanical Properties in Metal Additive Manufacturing. *NPJ Comput. Mater.* **2021**, *7*, 86. [CrossRef]
44. Long, H.; Lin, H.; Li, S.; Bai, Y.; Qin, L.; Xiao, T.; Qin, A. Nanomechanical Properties of CH_4-Containing Coal during CO_2 Storage under Different Injection Pressures Based on Molecule Dynamics. *Appl. Surf. Sci.* **2022**, *590*, 153126. [CrossRef]
45. Parsajoo, M.; Armaghani, D.J.; Mohammed, A.S.; Khari, M.; Jahandari, S. Tensile Strength Prediction of Rock Material Using Non-Destructive Tests: A Comparative Intelligent Study. *Transp. Geotech.* **2021**, *31*, 100652. [CrossRef]
46. Subramanian, S.M. Mechanical Properties of Materials: Definition, Testing and Application. *Int. J. Mod. Stud. Mech. Eng.* **2020**, *6*, 28–38. [CrossRef]
47. Raj, R.; Dixit, A.R.; Łukaszewski, K.; Wichniarek, R.; Rybarczyk, J.; Kuczko, W.; Górski, F. Numerical and Experimental Mechanical Analysis of Additively Manufactured Ankle–Foot Orthoses. *Materials* **2022**, *15*, 6130. [CrossRef]

polymers

MDPI

Article

Evaluation of U-Notch and V-Notch Geometries on the Mechanical Behavior of PVDF: The DIC Technique and FEA Approach

Ingrid C. S. Pereira [1], José Renato M. de Sousa [2] and Celio A. Costa [1,*]

1 Department of Metallurgical and Materials Engineering, COPPE/UFRJ, Federal University of Rio de Janeiro, Rio de Janeiro 21941-596, Brazil; ingrid_13@metalmat.ufrj.br
2 Department of Civil Engineering, COPPE/UFRJ, Federal University of Rio de Janeiro, Rio de Janeiro 21941-596, Brazil; jrenato@laceo.coppe.ufrj.br
* Correspondence: celio@metalmat.ufrj.br

Abstract: The notch effect of semicrystalline PVDF was investigated using U- and V-notch geometries with different depths, and tensile tests were performed at 23 °C using the DIC technique and FEA. Both unnotched and notched dumbbell-shaped specimens were subjected to tensile loading with the DIC technique to obtain mechanical curves and strain maps. The experimental data were compared to a numerical model, analyzing both global mechanical curves and local strain maps around the notch region to assess the accuracy of the simulations. The results demonstrated that the geometry and depth of the notch influence the mechanical behavior of PVDF, presenting a decrease in load and displacement compared to unnotched specimens. This aspect was corroborated by strain maps, which showed the increase in the local strain around the notch tip. For FEA, the global analysis indicated a good correlation with experimental results, and the local analysis demonstrated a reasonable agreement in strain map results within 0.5 mm of the notch neighborhood. Overall, the DIC technique and FEA provided a reliable evaluation of notch behavior on the PVDF used as pressure sheaths with reasonable precision.

Keywords: polyvinylidene fluoride (PVDF); digital image correlation (DIC); finite element analysis (FEA); notch geometry

1. Introduction

Notches are widely recognized as sources of stress concentration and have been extensively studied in various materials, focusing on factors like their shape (U- and V-notches), radius of curvature, and depth [1,2]. Hence, their presence is crucial in the design, fabrication, operation, and life assessment of several mechanical components, mainly due to their adverse impact on the material's embrittlement factor, which favors crack nucleation and reductions in the component tensile strength and fatigue resistance [3].

Over the past few decades, studies have delved into the notch effect in polymers employing a variety of experimental methods, including holographic interferometry [4], photoelasticity [5], and digital image correlation (DIC) [6,7], as well as analytical/numerical solutions (i.e., fracture parameters [8], William's equations [6,8], and finite element analyses (FEA)) [8]. However, these studies have utilized nonstandard notches under specific elastic or elastoplastic monotonic conditions [9–11]. Indeed, there is currently no consensus in the literature regarding whether specimen and notch geometry is sustainable for accurately assessing the notch effect in the mechanical behavior of polymers.

In recent years, applying the DIC technique and FE models has gained popularity in studying the mechanical behavior and the strain and stress fields related to the notch effect [12,13]. For example, Liu et al. [14] implemented the DIC method to measure deformation maps in notched hydrogels under tensile load, and the experimental results demonstrated a good agreement with an FE model. Similarly, Torabi et al. [15] and Bahrami et al. [16] assessed the fracture behavior of notched polymethyl–methacrylate (PMMA) using the DIC technique and FE analysis. Both studies reported a good correlation between results.

Understanding the strains or stresses at the notch tip is essential in evaluating the mechanical behavior of notches in polymers applied in advanced industries, such as the offshore oil and gas industry [17,18]. Nonetheless, the current notch tests are limited to amorphous polymers that exhibit brittle and quasi-brittle fractures below glass transition temperature (T_g), where the stress distribution field around the notch can be illustrated through equations that provide a linear elastic stress–strain relation [8,16,19]. As a result, a limited number of works use the DIC method to extract displacement/strain maps of the notch in semicrystalline polymers. In particular, semicrystalline polyvinylidene fluoride (PVDF) layers are often employed as pressure sheaths for offshore flexible pipes [20]. These pipes comprise concentric metallic and polymeric layers. The metallic layers resist the imposed mechanical loads, while the polymeric layers ensure fluid tightness and/or reduce friction and wear between the metallic layers. Regarding the polymeric layers, the pressure sheath internally seals the pipe, ensuring that the transported fluid leaks to the outer water environment [18].

A study conducted by Hund et al. [21] investigated PVDF using CNBR geometry and the DIC method to predict voids during tensile loading. It was observed that yielding was influenced by the stress triaxiality effect caused by a dual notch of CNBR geometry. Moreover, Ychisawa [22] applied the DIC technique to evaluate blunt notches made in the PVDF (S_{pb} method) but encountered difficulties creating displacement maps due to the complexity generated by significant deformation neighborhood defects. Pereira [9] studied the effect of two types of notches and cracks using strain maps during tensile tests from commercial grades of PVDF to analyze large deformations and utilized finite element analysis (FEA) for modeling purposes.

The present research aims to investigate the impact of U-notch and V-notch geometries on stress concentration in a commercial-grade PVDF using DIC and FEA. For the first time, dumbbell-shaped PVDF specimens with different notch depths were subjected to tensile tests at room temperature, and strain maps around the notch tip and tensile curves were obtained. Finally, the experimental results from the tensile tests were compared with those from numerical simulations to evaluate the mechanical behavior of notched PVDF specimens.

2. Materials and Methods

2.1. Experimental Methodology

2.1.1. Material and Specimen Preparation

The material studied was the PVDF with 3% plasticizer. Dumbbell-shaped (DBS) specimens with a constant thickness of 6 mm were machined in the longitudinal direction from commercially extruded pipes based on ISO 527-2 (Type 5A) [23]. Unnotched samples were compared to U-notch samples (curvature radii of 1 mm) and V-notch samples (curvature radii of 0.25 mm, and internal angle of 60°) with varying notch depths (0.2 mm, 0.6 mm, 1.0 mm, and 1.5 mm) to evaluate the notch effect. Figure 1 and Table 1 illustrate the specimens' notch geometries and test matrix, respectively.

The specimens were coated with black spray paint to create a random speckle pattern, allowing DIC measurements during the tensile tests. This coating has a unique speckle pattern that contrasts the black dots and the natural white surface of PVDF, facilitating the digital sensor's recognition of surface displacement during load application. The black covered area fraction Fr (%) and the speckle size through Ferret diameter D_F were measured using ImageJ [24] software (version 1.54d) to evaluate the quality of the speckle pattern. These values were also compared to the image processing of painted samples presented in a previous study [9].

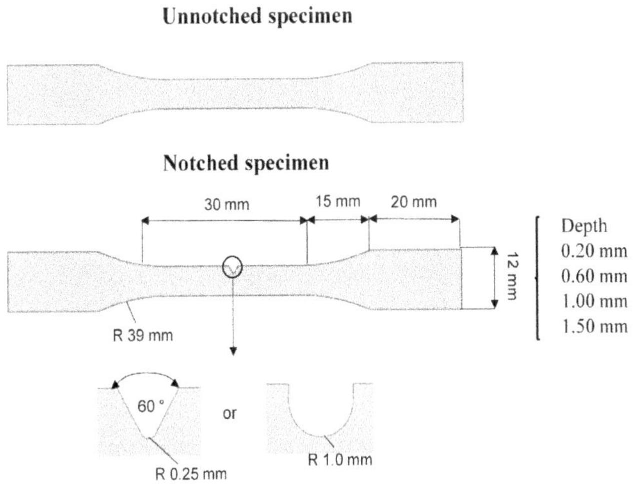

Figure 1. Illustration of unnotched and notched specimens.

Table 1. Notch groups and correspondent depth.

Specimens' Group	Depth (mm)
Unnotched	---
V-notch	0.2
	0.6
	1.0
	1.5
U-notch	0.2
	0.6
	1.0
	1.5

2.1.2. Experimental Setup of DIC

Five samples were tested per specimen condition to verify the reproducibility of the results obtained by mechanical tests. Uniaxial tensile tests combined with DIC were conducted until specimen fracture at room temperature (23 °C) under 5 mm/min displacement control, employing an electromechanical universal testing machine (INSTRON 5567) with a 10 kN load cell. Thus, experimental data were reaction forces measured by the load cell, displacements obtained from tensile control, and the strain maps plotted using the DIC system. For comparison, tensile tests were also conducted using a 25 mm extensometer to assess the accuracy of the DIC technique.

Throughout the experiment, optical images of the specimens' side surfaces were captured using a digital camera equipped with a 5MP resolution and a 100 mm Tokina lens. A light device and a computer were utilized to ensure image registration. The eCorr lab [25] acquisition software facilitated image acquisition at a speed of five frames per second (5Hz), with each DIC acquisition frame correlated with the load and displacement obtained by Blue Hill 3. The camera was positioned at 1.30 m for the unnotched specimens and 0.27 m for the notched samples. The Ncorr v.1.2 software [26], which runs on MATLAB R 2020a, was used to post-process the images. Figure 2 displays the experimental setup employed for the acquisition of images (Step I) and post-processing (Step II) to obtain strain maps (Step III).

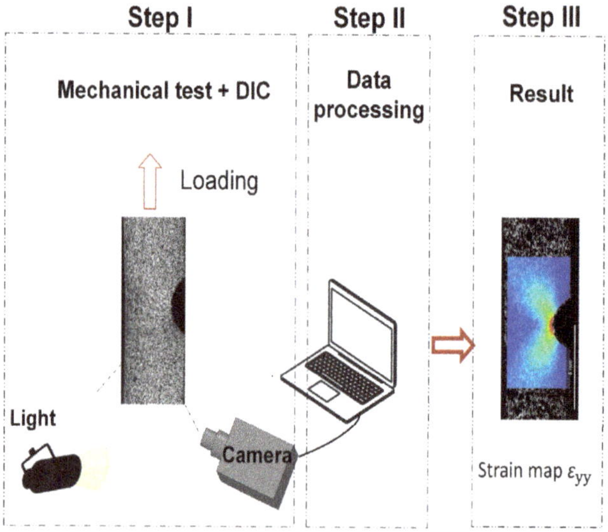

Figure 2. Schematic of 2D-DIC setup: Step I—acquisition of images, Step II—post-processing, and Step III—the resulting strain map.

2.1.3. FE Model Implementation

In this study, a finite element (FE) model was employed to simulate the behavior of notches in PVDF samples subjected to monotonic tensile tests. The FE model was two-dimensional (2D), representing the geometry of the tested samples, and was developed in ANSYS® [27]. As several FE meshes were employed in this study, their construction was automatized in a macro written in APDL (ANSYS Parametric Design Language), allowing their pre-processing, solution, and post-processing.

The notched tensile specimens of different depths were modeled with 8-node quadrangular plane stress finite elements, named PLANE183 in ANSYS®. Each node has two degrees of freedom, i.e., translations regarding the X and Y directions. Moreover, a triangular degenerated form with six nodes was employed in regions with irregular geometry. A view of a typical FE mesh is presented in Figure 3. PLANE183 has quadratic interpolation functions suitable for modeling regions with high-stress gradients, such as those found near notches, justifying its choice instead of elements with linear interpolation functions. For instance, Thirumump et al. [28] employed this element to evaluate stress concentrations in plates with holes, obtaining satisfactory results.

Figure 4 shows a close-up view of the mesh configuration used in the FE model. As a result of the high stress at the notches' tips, an unstructured refined mesh was adopted in the vicinity of the notch neighborhood (V-notch or U-notch), whereas a coarser structured FE mesh was constructed following the specimen geometry in regions far from the notches. Moreover, the red dashed line corresponds to the medium line on which the strain values were extracted for later comparison with DIC results.

Additionally, the constructed FE models aimed to simulate the failure of the PVDF samples. Hence, large strains and material nonlinearities were expected. Geometric nonlinearities, such as large strains, are directly accounted for in PLANE183. In contrast, this work assumed a rate-independent material response, i.e., viscoelastic effects were disregarded. Hence, the material constitutive model required defining a stress vs. strain relation by considering a yield (failure) criterion, a hardening rule, and a flow rule. The relation between the uniaxial stress state and the multiaxial stress state induced in the tests was established with the von Mises yield criterion, assuming PVDF as a ductile material. Moreover, the imposed load was monotonic, and isotropic hardening was considered, following the recommendation presented in ANSYS® (version 19.2) [27]. Finally, the flow rule was associative, i.e., the potential was identical to the von Mises yield surface.

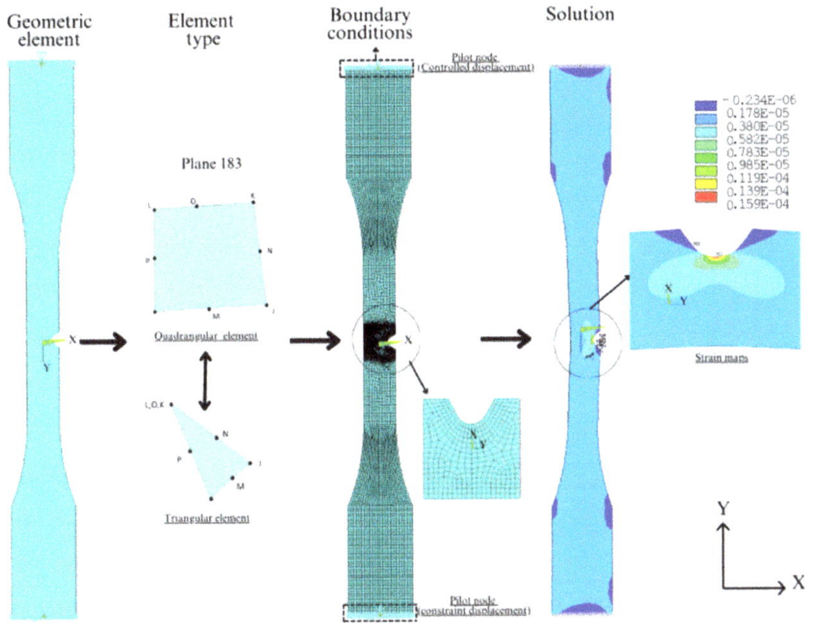

Figure 3. Element finite model showing the tensile test, mesh generation, and boundary conditions in the numerical simulations.

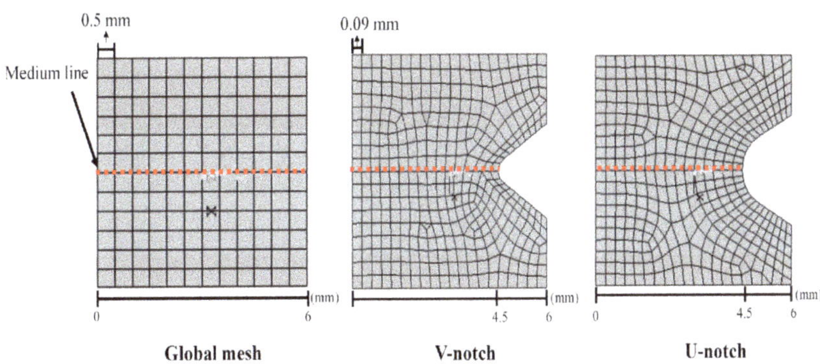

Figure 4. Illustration of a close-up to observe global mesh and mesh in the notch region.

Table 2 and Figure 5 present the input of material properties in the numerical model. The true stress–strain curve's components were calculated using Equations (1) and (2) [29], which were adjusted to the multilinear model from the stress–strain curve of the unnotched samples obtained by tensile tests with the DIC technique.

$$\sigma_{true} = \sigma_{eng}\left(\sqrt{2 \times \varepsilon_{yy-lag}} + 1\right) \quad (1)$$

$$\varepsilon_{true} = In\left(\sqrt{2 \times \varepsilon_{yy-lagg}} + 1\right) \quad (2)$$

where ε_{yy} is the Lagrangian strain in the longitudinal direction (Y direction).

Table 2. Properties of commercial grades of PVDF.

Properties	Values (Unity)
Density	1.78×10^5 kg/m^3
Poisson's ratio	0.43
Elastic modulus	1280.15 MPa

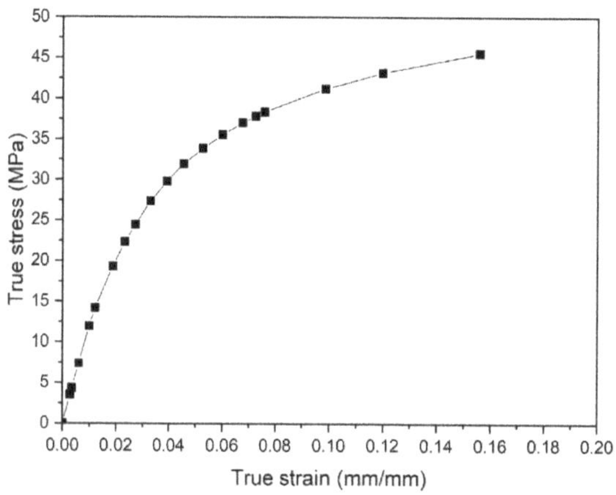

Figure 5. True stress–strain curve of PVDF adjusted to multilinear model.

To simulate the tensile test, the geometry and boundary conditions identical to those imposed on the experimental tensile tests were selected. The boundary conditions were applied at pilot nodes positioned at the ends of each model. The pilot nodes were meshed with CONTA175 elements and were rigidly connected to the nodes on the edge lines of the sample, which were meshed with TARGE170 elements. Hence, rigid contact pairs were formed, and a translation in the Y-axis direction was applied at the pilot node at one end of the specimen. The other end was constrained in the X and Y directions to avoid rigid body motion.

Furthermore, Pereira [9] conducted an FE mesh study to ensure the FE model's accuracy, indicating that a maximum element edge length of 0.5 mm was required for the structured region. In comparison, a maximum length of 0.09 mm was demanded for the notch region. Table 3 presents the number of elements and nodes for each type of notch geometry.

Table 3. The total number of elements and nodes corresponding to each notch geometry with different depths.

Specimens' Group	Depth (mm)	Number of Elements	Number of Nodes
V-notch	0.2	8622	29,161
	0.6	9087	27,550
	1.0	9047	27,510
	1.5	9003	27,288
U-notch	0.2	9450	28,794
	0.6	9017	23,760
	1.0	8847	26,860
	1.5	8414	25,547

Lastly, all numerical analyses were carried out using the sparse solver of ANSYS® [27], employing automated load steps. At each load step, the convergence was achieved if the

L2 norms of the residual forces, and moments were less than 0.1% of the absolute value of the acting total forces and moments. The first load step corresponded to 0.1% of the total imposed load, but depending on the convergence rate, this load step may be increased to 10% of the total imposed load or reduced to 0.01%. Considering all these aspects, a typical FE analysis was concluded after 2.1 h in an Intel i7 CPU and 16 GB of RAM.

3. Results and Discussion

3.1. Tensile Behavior of Unnotched Specimens: DIC and Extensometer

Mechanical tests of the unnotched specimens of PVDF were performed to obtain material properties for the FE model and investigate the accuracy and reliability of the DIC technique, as described in Section 2.1.2. The tests' results revealed that the stress–strain curves generated with both methods were almost identical until the yield point, as seen in Figure 6, indicating a good agreement considering the expected viscoelastic behavior of semicrystalline polymers [30]. The yield strain measured from stress–strain curves using the extensometer was 14.5 ± 0.21%, while with the DIC method, it measured 16.2 ± 0.29%. This slight discrepancy can be attributed to two factors. Firstly, the extensometer, due to being a contact technique, may have introduced an unwanted stress concentration at the surface of the specimen, thus predicting the necking nucleation. Secondly, despite its efficiency in obtaining reliable and accurate strain data, the accuracy of the DIC in measuring strain values depends on several factors, such as the speckle pattern quality, the experimental setup, and the degree of noise during loading, which may affect the accuracy of the partial derivatives functions applied to calculate the strains, according to Blabber, Adair, and Antoniou [26].

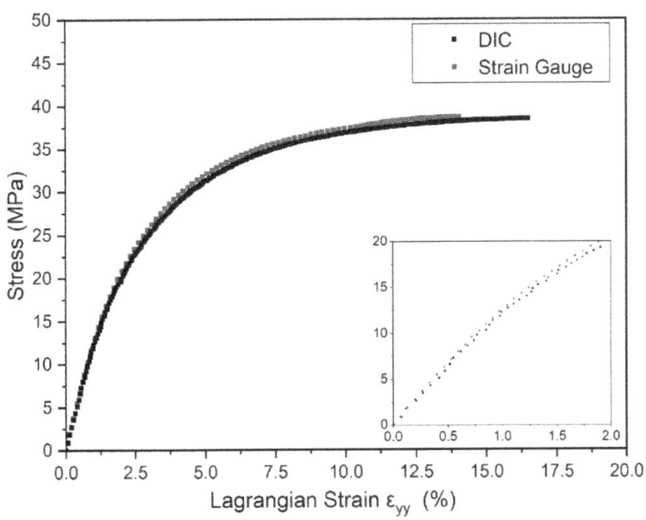

Figure 6. Comparison between stress versus Lagrangian strain εyy curves of DIC and strain gauge at 23 °C under the rate of 5 mm/min of unnotched specimens.

The material's mechanical response was analyzed through tensile tests using DIC's images without post-processing. In Figure 7, the load values associated with local maximum strain at the notch tip are presented as follows: Point (A) at 1.2%, Point (B) at 3.5% (related to flexible pipe's dynamic operation [31]), Point (C) at 7.0% (related to flexible pipe's static operation [31]), and Point (D) at 16.2% (corresponding to the yield point criterium). Figure 7a shows that the material exhibited an apparent linear elastic behavior until the yield point (Point (D)), followed by a decrease in stress from around 38.9 MPa to 30.5 MPa at Point (E) once necking occurred. The stress remained constant (30.5 MPa) during the neck propagation. The PVDF sample demonstrated a ductile behavior associated with yielding and neck nucleation with the whitening phenomenon, which has also been reported by Defebvin et al. [32] and Castagnet et al. [33,34]. Figure 7b shows this phenomenon in the neck region, indicating a cavitation process followed by void coalescence in the samples. The whitening presented at unnotched samples led to the contrast loss of speckle pattern in the surface of the specimens under tensile test using the DIC at high strains. To ensure reliable and accurate data, the strain maps extracted until the yield point ($d\sigma/d\varepsilon = 0$, necking locus [35]) were compared, as illustrated until Point (D).

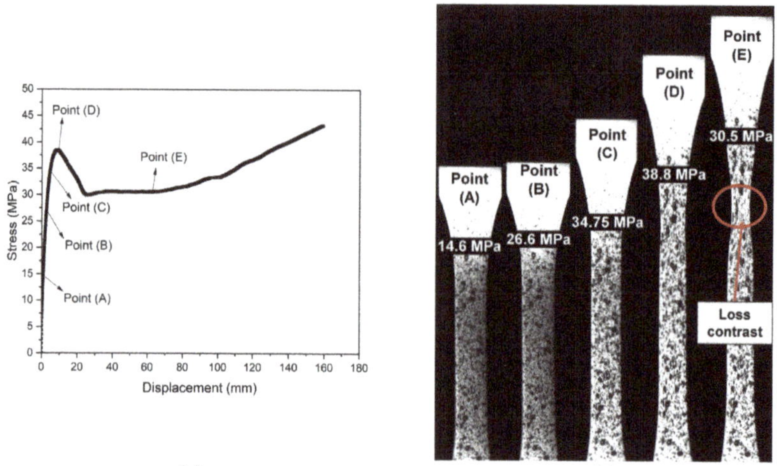

(**a**) (**b**)

Figure 7. Illustrations (**a**) stress versus time with (**b**) images during tensile test using DIC to specific points: (A) stress at the linear–elastic region, (B) stress at 3.5% of strain, (C) stress at 7.0% of strain, (D) stress at the yield point, and (E) stress at necking.

3.2. Strain Map Distribution for Unnotched Specimens

One of the significant advantages of the DIC is the ability to generate strain maps on the specimen surface. Figure 8 shows the typical strain maps obtained from the DIC for specific stresses in the stress–time curve shown in Figure 7a. At Point (A), an increase in the deformation is observed, with a homogenous Lagrangian strain in the Y-axis (ε_{yy}^{Lag}) distribution in the region of interest (ROI), i.e., below 1.2%. The values observed in the strain maps when PVDF reached 3.5% are noteworthy (27.2 ± 0.4 MPa, Point (B)) and 7.0% (34.6 ± 0.3 MPa, Point (C)) strains, which are associated with the limit strains for PVDF operation under dynamic and static operation as a pressure sheath according to API SP 17J [31], respectively. Point (D) corresponds to the yield point, which, on average, corresponds to a stress of 38.9 ± 0.5 MPa and a strain of 16.2 ± 0.29%, where the red mark shows the area that nucleated the neck. The DIC accurately measured the developed large strains.

The DIC technique also enabled measuring the deformation in both longitudinal and transversal directions until the yield point, as shown in Figure 9, indicating a negative ε_{xx}^{Lag} due to the reduction in the transversal section and an increase in the longitudinal section ($\varepsilon_{yy}^{Lag} > 0$), as expected in tensile tests. Furthermore, as the components ε_{yy}^{Lag} and ε_{xx}^{Lag} were also obtained in the elastic region, the Poisson coefficient v could also be determined. At a strain rate of 5 mm/min and a temperature of 23 °C, the mean value of v was found to be 0.43 ± 0.04. It is important to note that this average value is only valid for the specific test conditions because v is a time-dependent property for polymers, opposing what is observed in metallic materials [36]. Compared to the literature, the values of v obtained for PVDF (0.43 ± 0.04) were similar to those reported by Castagnet et al. [34], i.e., 0.36 to 0.47.

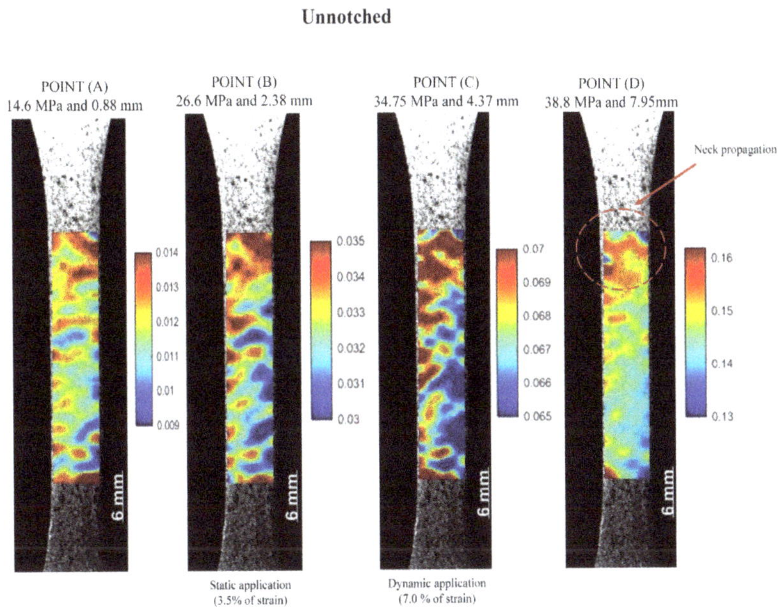

Figure 8. Lagrangian strain εyy maps of unnotched specimens that correspond to specific points of Stress versus Lagrangian strain εyy curve: (A) stress at the linear–elastic region, (B) stress at 3.5% of strain, (C) stress at 7.0% of strain, and (D) stress at the yield point.

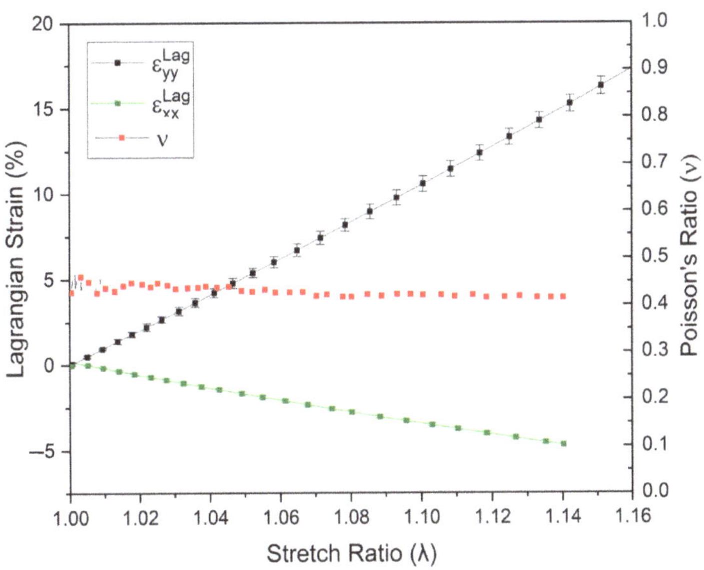

Figure 9. Poisson's ratio (v) and longitudinal (ε_{yy}^{Lag}) and transversal (ε_{xx}^{Lag}) strains obtained using the DIC at different stretch levels until the yield point for the unnotched PVDF specimen.

3.3. Tensile Behavior of Notched Specimens

Tensile tests were performed on the notched specimens of PVDF under the same conditions as the unnotched tests. Figures 10 and 11 show the measured load vs. displacement curves, evidencing nonlinear responses in all the notched specimens. The curves obtained for the notched specimens were close to the unnotched ones up to the associated failure loads. However, the failure loads of the notched specimens were lower than those observed in the unnotched specimens, and a significant decrease was also observed with an increase in notch depths. The failure of the notched specimens is related to stress concentrations around the notch root, causing local plastic deformations and the rupture of the specimens. As a result, the PVDF ductility perceived in the unnotched tests was significantly reduced, as observed, e.g., in polyolefins [37].

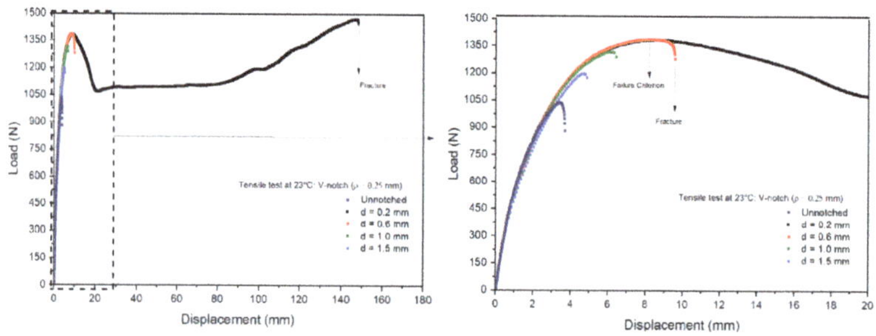

Figure 10. Load–displacement curves for V-notch specimens.

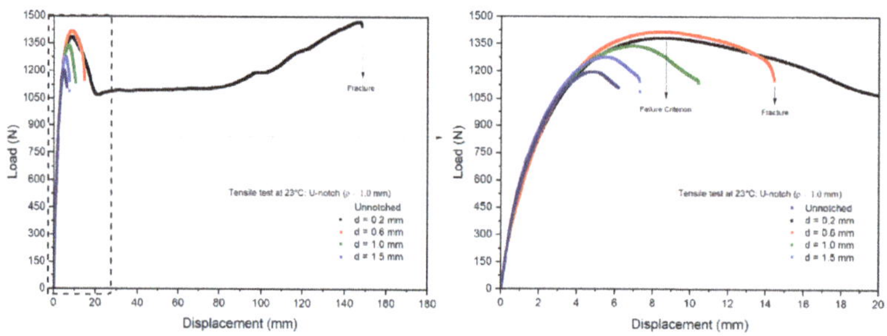

Figure 11. Load–displacement curves for U-notch specimens.

Table 4 presents the average maximum loads and displacements measured in the notched specimens' tests normalized by the unnotched values. First, the results indicated good repeatability with small standard deviations, thus leading to coefficients of variation lower than 5.0%. For a notch depth of 0.2 mm, the V-notch samples exhibited values similar to the unnotched samples, while the U-notch samples with a depth of 0.2 mm had slightly higher values than the unnotched samples, possibly due to a local hardening effect. The loads and displacements decreased linearly for notch depths higher than 0.6 mm. The U-notch load drop was less steep than observed in the V-notch specimens, as the stress concentration increased at the notch root with decreasing curvature radii [38].

Table 4. Maximum load and displacement to each notch geometry with different depths.

Specimens' Group	Depth (mm)	Maximum Load (MPa)	Maximum Displacement (mm)
Unnotched		1393.0 ± 31.4	8.0 ± 0.2
V-notch	0.2	1391.4 ± 5.7	8.4 ± 0.2
	0.6	1305.4 ± 30.6	6.1 ± 0.4
	1.0	1197.3 ± 6.9	4.5 ± 0.2
	1.5	1042.1 ± 17.2	3.3 ± 0.1
U-notch	0.2	1447.0 ± 19.5	8.3 ± 0.3
	0.6	1354.0 ± 19.4	7.2 ± 0.3
	1.0	1263.4 ± 23.4	5.6 ± 0.1
	1.5	1184.1 ± 21.5	4.7 ± 0.2

Furthermore, the impact of notches on the tensile properties can be understood by examining the PVDF micro-mechanisms. Shear flow or yielding and cavitation processes, which compete during the tensile tests, influence the fracture of PVDF above *Tg* [33,34]. The yielding process occurs uniformly in unnotched samples and is simultaneous to the whitening onset. On the other hand, the notches significantly alter the mechanical response of the samples due to the modified local stress state. Significant hydrostatic stress is ahead of the notch when cavitation tension reaches a critical value. As a result, notching constrains plastic yielding, and elastic performance and cavitation processes dominate at the notch root. Moreover, stress whitening around the notches is caused by these cavities, similar to what was observed in other PVDF specimens [39,40].

3.4. Strain Maps for Notched Specimens Using the DIC Technique

The DIC analysis was conducted in the notched specimens to assess the nucleation of the plastic zone at the notch tip, where damage was presumed to be significant. For comparison purposes in this study, the reference points are the same used in Figure 8. Then, the maximum strain measured was carried out up to the strain near the notch region to reach the yield strain to the unnotched specimens (Point (D) ~16%).

Figures 12 and 13 present the strain maps for depths of 0.2 mm and 1.5 mm in V- and U-notch geometries, respectively, as examples. There is a clear tendency toward strain accumulation in the notch region with the surrounding material at lower deformation, as demonstrated when comparing Point (A) to Point (B), where localized deformation increased by around 2.3%. Additionally, considering API 17J [31], the limit strain in the pressure sheath of a flexible pipe (dynamic operation) is 3.5%, and for a flexible flowline (static operation), it is 7.0%. These values correspond to the maximum deformations in the notch front at Point (B) and Point (C), respectively, and were observed before reaching the maximum load in all the notched conditions.

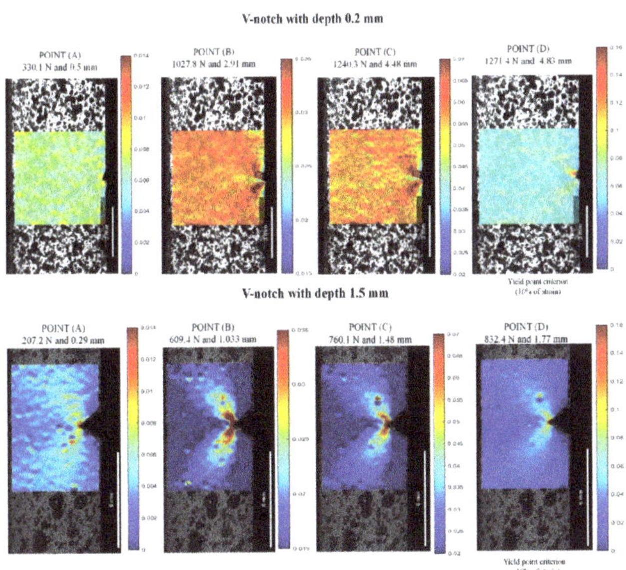

Figure 12. Lagrangian strains ε_{yy} in V-notch tests at 23 °C and 0.2 and 1.5 mm depths: (A) stress at the linear–elastic region, (B) stress at 3.5% of strain, (C) stress at 7.0% of strain, and (D) stress at the yield point.

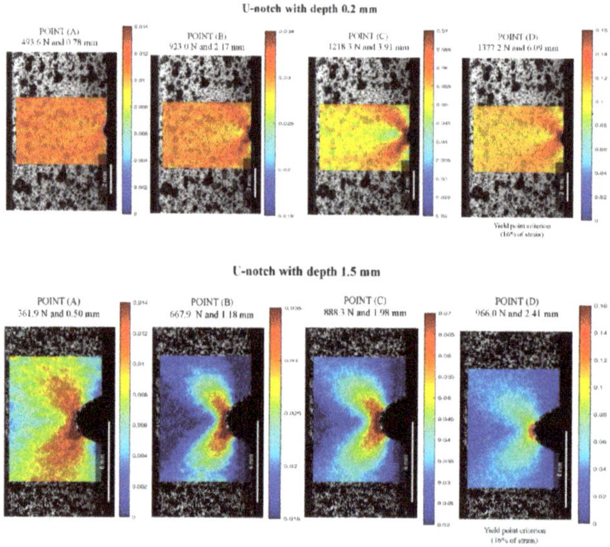

Figure 13. Lagrangian strains ε_{yy} in U-notch tests at 23 °C and 0.2 and 1.5 mm depths: (A) stress at the linear–elastic region, (B) stress at 3.5% of strain, (C) stress at 7.0% of strain, and (D) stress at the yield point.

As illustrated in the strain maps for the V-notch sample (Figure 12), increasing the notch from 0.2 to 1.5 mm led to a significant intensification of local stress concentration compared to the unnotched specimens. At a depth of 0.2 mm, the specimen presented a discrete local stress concentration, indicating displacements of 2.91 mm at 1027.8 N for Point (B) and 2.83 mm at 1240.3 N for Point (C). On the other hand, as the depth increased to 1.5 mm, the strain at the notch tip showed a local strain more spread out from the notch with a butterfly wing format (BWF), resulting in a reduction of around 35.0% for displacement and 59.5% for load, concerning 3.5% (Point (B)) and 7.0% (Point (C)). The increase in local strain concentration occurs because the elastic bulk material severely constrains the plastic deformation zone ahead of the notch for all strain fields. As a result, the area far from the defect zone experiences lower and more uniform deformation values [1,9]. This pattern was also observed in the U-notch specimens in Figure 13, where there was around a 49.3% decrease in displacement and approximately 27.6% decrease in load at Point (B) and Point (C) from 0.2 to 1.5 mm depth.

As depicted in the strain maps for the U-notch (Figure 13), different notch geometries induced a distinct local strain concentration profile, thereby altering the mechanical behavior of PVDF, as seen in Figures 10 and 11. U-notch specimens presented a damage zone spread far into the bulk compared with V-notch specimens, and V-notch tip blunting was limited. Indeed, for notches with the same depth, a smaller curvature radius at the notch root resulted in a lower load needed to achieve the same strain. For instance, at a depth of 0.2 mm and 1.2% strain, a V-notch required a load of 330.1 N, while a U-notch required 493.6 N. In this case, deformation was evenly distributed in both areas. As the load increased from Point (B) to Point (D), the V-notch concentrated all the strain at its tiny tip, while the U-notch was able to spread the deformation zone in a BWF. As shown in Figure 12, for a 1.5 mm depth, the V-notch concentrated the highest strain at its tip up to Point (C), and above it, the highest deformation zone spread from the tip in a BWF. On the other hand, as shown in Figure 13, the U-notch with 1.5 mm depth spread the BWF deformation zone up to Point (C) and then concentrated it at its tip, showing an opposite behavior.

3.5. FEA of Notched Specimens

The modeling results of the PVDF notches' behavior using the finite element analysis (FEA) were compared to the experimental results to verify the proposed numerical model's suitability and accuracy.

Figures 14 and 15 and Table 5 compare the experimental and simulated load–displacement curves for both types of notches at all depths and considering the points indicated in Figure 8. The numerical model well captured the global behavior of the notched specimens. The numerical curves followed the experimental ones until reaching a plateau, indicating the samples' rupture. As the notch depth increased, the displacement associated with the beginning of the plateau was also reduced. For instance, V- or U-notches with a 0.2 mm depth had a plateau starting around 4.6 mm displacement, while for a 1.5 mm depth, the plateau began at about a 3.0 mm displacement. The difference between the experimental and the FE results during failure may be attributed to how the defect (notch) develops and propagates during the experimental tests and the FE analyses. In the experimental tests, the defect propagated and tore the samples with the increasing imposed loads. Instead, the FE mesh remained continuous throughout the analysis, and no gaps (or voids) were initiated. Hence, the actual rupture of the material, which would demand the discontinuity of the FE mesh, was not modeled. Still, the increasing acting strains progressively reduced the element stiffness to simulate the sample's failure. This approach modifies the stress field close to the notch tip, particularly given the constraint caused by a sharp curvature radius, possibly altering the failure mechanism development. It is recognized that this is a limitation of the proposed approach that should be investigated in future studies.

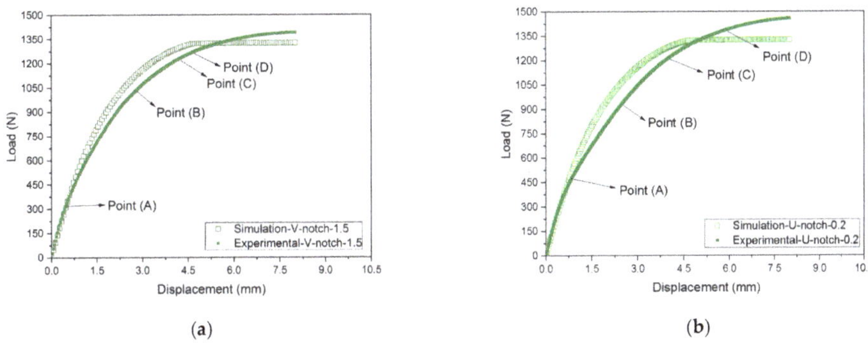

Figure 14. Comparison between experimental and FE responses considering notches with 0.2 mm depth in (**a**) V-notch and (**b**) U-notch samples: (A) stress at the linear–elastic region, (B) stress at 3.5% of strain, (C) stress at 7.0% of strain, and (D) stress at the yield point.

Table 5 compares the loads related to Points (A) to (D) obtained in the experimental tests and those estimated with the FE model considering the V- and U-notches and different depths. Regarding the V-notch, the experimental forces were about 3.4% higher on average, with a standard deviation of 10%, leading to a coefficient of variation of 9.7%. Moreover, the U-notch results indicate experimental values of only 0.1% higher than the FE ones on average, with a standard deviation of 6.9%, resulting in a coefficient of variation equal to 6.9%. Hence, despite some significant specific differences, the FE model results agreed well with the experimental measurements on average.

In the prediction of localized deformations, Lagrangian strains were extracted along the Y-axis from the notch tip for both DIC and FEA, as illustrated in Figure 4. Figures 16–19 show the corresponding results at Point A (the load associated with the elastic region) and Point D (the load related to the yield point criterium), in which the region near the notch tip presents values of higher deformation.

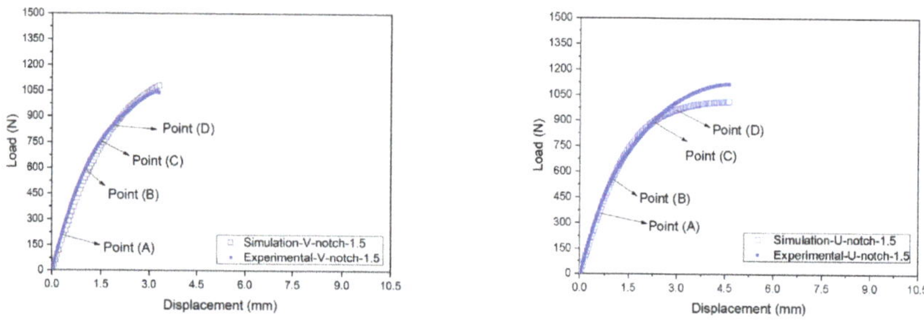

(**a**) (**b**)

Figure 15. Comparison between experimental and FE responses considering notches with 1.5 mm depth in (**a**) V-notch and (**b**) U-notch samples: (A) stress at the linear–elastic region, (B) stress at 3.5% of strain, (C) stress at 7.0% of strain, and (D) stress at the yield point.

Table 5. Comparison of load of notched specimens between FEA-simulated and experimental (Exp.) data.

Specimens' Group	Depth (mm)	Load (N)							
		Point (A)		Point (B)		Point (C)		Point (D)	
		Exp.	FEA	Exp.	FEA	Exp.	FEA	Exp.	FEA
V-notch	0.2	330.1	279.2	1027.8	1124.5	1240.3	1305.4	1271.4	1322.1
	0.6	380.9	395.8	760.6	881.3	986.4	1104.2	1187.2	1222.1
	1.0	231.7	221.7	500.2	551.5	708.7	800.5	1042.6	1087.8
	1.5	207.2	169.8	609.4	556.2	760.1	732.4	832.4	818.5
U-notch	0.2	493.6	540.9	923.0	1064.6	1218.3	1276.5	1377.2	1325.5
	0.6	427.7	433.3	817.3	937.2	1069.4	1170.2	1248.3	1225.4
	1.0	399.6	396.2	799.9	794.5	1023.7	1022.9	1173.8	1123.6
	1.5	361.9	340.1	667.9	675.0	888.3	891.0	966.0	940.7

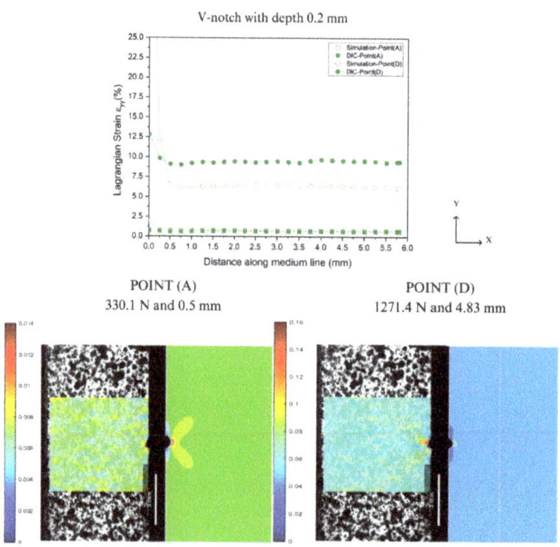

Figure 16. Comparison between DIC technique and FEA at Point A and Point D for V-notch tip with 0.2 mm depth.

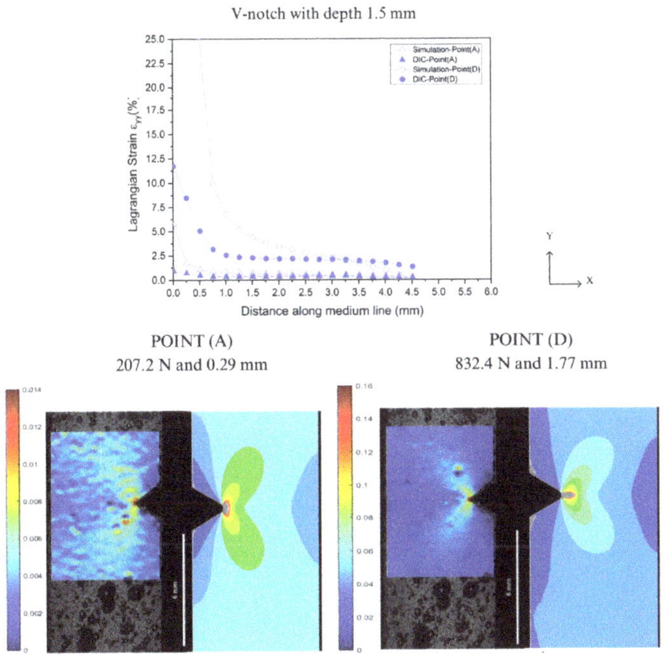

Figure 17. Comparison between DIC technique and FEA at Point A and Point D for V-notch tip with 1.5 mm depth.

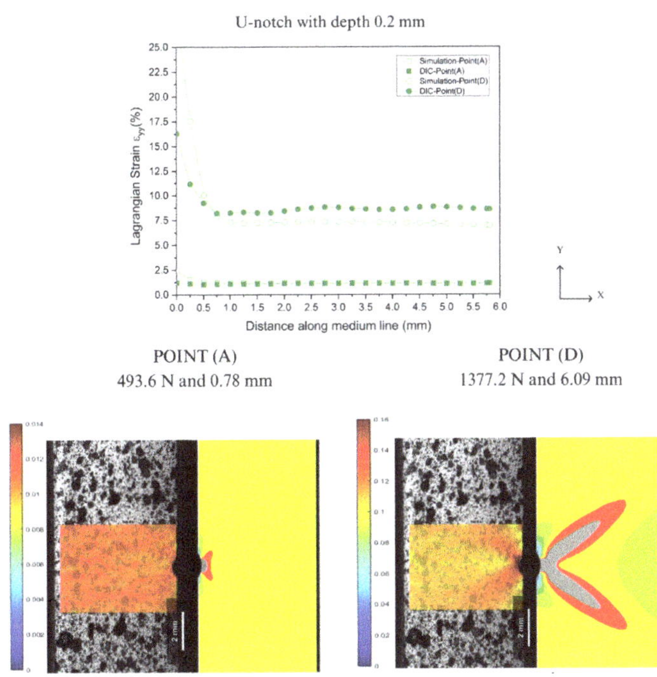

Figure 18. Comparison between DIC technique and FEA at Point A and Point D for U-notch tip with 0.2 mm depth.

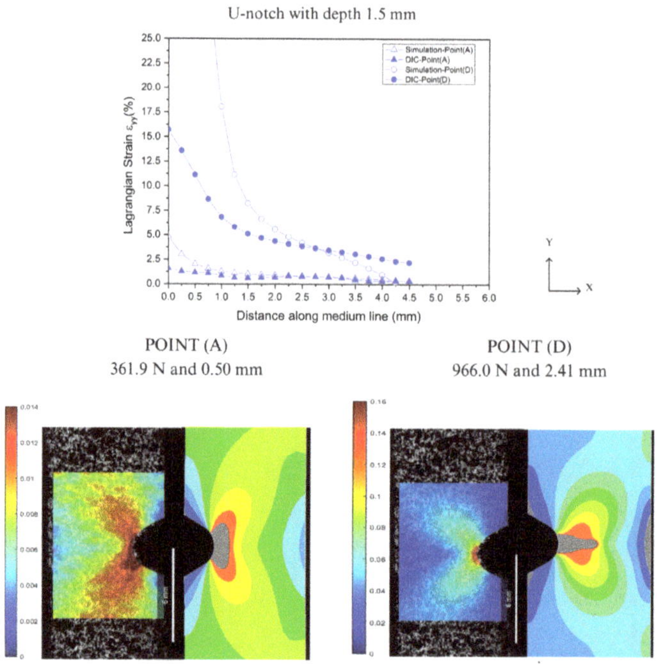

Figure 19. Comparison between DIC technique and FEA at Point A and Point D for U-notch tip with 1.5 mm depth.

In Figures 16–19, the accuracy and precision of the strain maps near the edge of the stress concentrators (below 0.5 mm) are more sensitive to modifying DIC parameters for ductile materials. This observation is consistent with Liu's report on the assessment of notches [41]. DIC results can be significantly impacted by the quality of speckle patterns, resolution, and discontinuity of the mesh. Moreover, the localized whitening phenomenon at the notch tip, which is usual in PVDF [33], plays an important role in the resolution levels used in FEA and DIC because the numerical model used in the analysis did not consider the whitening effect and viscoelastic behavior typical of PVDF. On the other hand, as the observation point moved away from the notch edge toward Point A (at 0.5 mm), both methods showed good convergence with only a slight difference in results.

4. Conclusions

PVDF was machined with U- and V-notches considering four different depths (0.2, 0.6, 1.0, and 1.5 mm). The samples were subjected to tensile tests at 23 °C, and the DIC technique was employed to map the related strains until the samples' failure. Then, the mechanical (strain gauges) and DIC results were analyzed and compared to the FE models that reproduced the experimental tests, leading to the following conclusions:

- In the experimental tests, the decay of the failure loads was observed with an increase in notch depths and a reduction in the notch radius in both U and V geometries. The decay is associated with an increased three-dimensional stress state at the notch tip.
- The DIC method showed the notch effect on the samples' deformation by generating strain maps. Severe constraints were detected, and the deformation around the notch tip under tensile loading was quantified.
- The FEA showed similar load vs. displacement curves compared to those obtained experimentally. Hence, the FE model well captured the global behavior of the unnotched and notched samples.

- When comparing the DIC technique and FEA, it was observed that numerical simulation can lead to deviations, possibly due to the continuous approach adopted in the FE models, which may affect the stress field close to the notches. This aspect may be studied in future work. Additionally, the difference in resolution levels between the DIC technique and FEA can influence the results measured for strain around the notch. However, a reasonable convergence between numerical and experimental strain results was observed at points distant 0.5 mm or higher from the defect.

Author Contributions: Conceptualization, I.C.S.P. and C.A.C.; methodology, I.C.S.P. and J.R.M.d.S.; software, I.C.S.P. and J.R.M.d.S.; validation, I.C.S.P.; formal analysis, I.C.S.P.; investigation, I.C.S.P., J.R.M.d.S. and C.A.C.; resources, C.A.C.; data curation, I.C.S.P.; writing—original draft preparation, I.C.S.P.; writing—review and editing, J.R.M.d.S. and C.A.C.; visualization, C.A.C.; supervision, J.R.M.d.S. and C.A.C.; project administration, C.A.C.; funding acquisition, C.A.C. All authors have read and agreed to the published version of the manuscript.

Funding: This study was financed in part by the Coordenação de Aperfeiçoamento de Pessoal de Nível Superior—Brasil (CAPES)—Finance Code 001, Shell Brasil Petróleo Ltda via ANP R&D Levy Clause and CNPq (Conselho Nacional de Desenvolvimento Científico e Tecnológico).

Institutional Review Board Statement: Not applicable.

Data Availability Statement: Data are contained within the article.

Conflicts of Interest: The authors declare no conflicts of interest.

References

1. Hertzberg, R.W. *Deformation and Fracture Mechanics of Engineering Materials*, 4th ed.; John Wiley & Sons: New York, NY, USA, 1992; Volume 9.
2. Schijve, J. Stress Concentration at Notches. In *Fatigue of structure and Materials*; Springer: Berlin/Heidelberg, Germany, 2008; pp. 59–88.
3. Qu, R.; Zhang, P.; Zhang, Z. Notch Effect of Materials: Strengthening or Weakening? *J. Mater. Sci. Technol.* **2014**, *30*, 599–608. [CrossRef]
4. Peterson, T.L.; Ast, D.G.; Kramer, E.J. Holographic Interferometry of Crazes in Polycarbonate. *J. Appl. Phys.* **1974**, *45*, 4220–4228. [CrossRef]
5. Ayatollahi, M.R.; Mirsayar, M.M.; Dehghany, M. Experimental Determination of Stress Field Parameters in Bi-Material Notches Using Photoelasticity. *Mater. Des.* **2011**, *32*, 4901–4908. [CrossRef]
6. Abshirini, M.; Dehnavi, M.Y.; Beni, M.A.; Soltani, N. Interaction of Two Parallel U-Notches with Tip Cracks in PMMA Plates under Tension Using Digital Image Correlation. *Theor. Appl. Fract. Mech.* **2014**, *70*, 75–82. [CrossRef]
7. Dehnavi, M.Y.R.; Eshraghi, I.; Soltani, N. Investigation of Fracture Parameters of Edge V-Notches in a Polymer Material Using Digital Image Correlation. *Polym. Test.* **2013**, *32*, 778–784. [CrossRef]
8. Ayatollahi, M.R.; Torabi, A.R.; Rahimi, A.S. Brittle Fracture Assessment of Engineering Components in the Presence of Notches: A Review. *Fatigue Fract. Eng. Mater. Struct.* **2015**, *39*, 267–291. [CrossRef]
9. Pereira, I.C.S. *Crack and Notch Sensitivity of Polyvinylidene Fluoride (PVDF) Experimentally Evaluated by DIC and Compared to FEM*; UFRJ: Rio de Janeiro, Brazil, 2021.
10. Bhadori, R.; Ayatollahi, M.R.; Shahmohammadi, M.; Berto, F. Geometry Effects on Mode I Brittle Fracture in U-Notched Specimens. *Fatigue Fract. Eng. Mater. Struct.* **2021**, *44*, 901–915. [CrossRef]
11. Gómez, F.J.; Elices, M. Fracture of Components with V-Shaped Notches. *Eng. Fract. Mech.* **2003**, *70*, 1913–1927. [CrossRef]
12. Brugo, T.M.; Campione, I.; Minak, G. Investigation by Digital Image Correlation of Mixed-Mode I and II Fracture Behavior of Polymeric IASCB Specimens with Additive Manufactured Crack-like Notch. *Materials* **2021**, *14*, 1084. [CrossRef]
13. Spagnoli, A.; Terzano, M.; Brighenti, R.; Artoni, F.; Carpinteri, A. How Soft Polymers Cope with Cracks and Notches. *Appl. Sci.* **2019**, *9*, 1086. [CrossRef]
14. Liu, M.; Guo, J.; Hui, C.Y.; Zehnder, A.T. Application of Digital Image Correlation (DIC) to the Measurement of Strain Concentration of a PVA Dual-Crosslink Hydrogel Under Large Deformation. *Exp. Mech.* **2019**, *59*, 1021–1032. [CrossRef]
15. Torabi, A.R.; Bahrami, B.; Ayatollahi, M.R. On the Use of Digital Image Correlation Method for Determining the Stress Fi Eld at Blunt V-Notch Neighborhood. *Eng. Fract. Mech.* **2020**, *223*, 106768. [CrossRef]
16. Bahrami, B.; Ayatollahi, M.R.; Torabi, A.R. In-Situ Brittle Fracture Analysis of Sharp V-Notched Components Using Digital Image Correlation. *Theor. Appl. Fract. Mech.* **2020**, *106*, 102484. [CrossRef]
17. Esaklul, K.A.; Mason, J. Nonmetallics Applications in Oil and Gas Production (Pipes, Liners, Rehabilitations). In *Trends in Oil and Gas Corrosion Research and Technologies: Production and Transmission*; Elsevier Ltd.: Amsterdam, The Netherlands, 2017; pp. 627–660. ISBN 9780081012192.

18. *API RP 17B*; Recommended Practice for Flexible Pipe. American Petroleum Institute: Washington, DC, USA, 2014.
19. Razavi, S.M.J.; Ayatollahi, M.R.; Shahmohammadi, M.; Berto, F. Geometry Effect on Fracture Behavior of V-Notched Specimens. *Procedia Struct. Integr.* **2020**, *26*, 234–239. [CrossRef]
20. Silva, A.J.d.J. *Avaliação Do Evelhecimento Do Poli (Fluoreto de Vinilideno) (PVDF) Visando Aplicações Em Estruturas Para Contato Com Etanol Combustível*; Universidade Federal do Rio de Janeiro: Rio de Janeiro, Brazil, 2017.
21. Hund, J.; Møgster, H.; Nordmark, S.; Henrik, P.; Johnsen, J.; Holm, A. Impact of Stress Triaxiality, Strain Rate, and Temperature on the Mechanical Response and Morphology of PVDF. *Polym. Test.* **2022**, *114*, 107717. [CrossRef]
22. Ychisawa, L.K. *Uso Do Método Spb Com Uso de Digital Image Correlation (DIC) Para Avaliação Da Tenacidade à Fratura de PVDF*; Universidade Federal do Rio de Janeiro: Rio de Janeiro, Brazil, 2019.
23. *ISO-527-2*; Plastics—Determination of Tensile Properties—Part 2: Test Conditions for Moulding and Extrusion Plastics 2012. International Organization for Standardization: Geneva, Switzerland, 2012.
24. Schneider, C.A.; Rasband, W.S.; Eliceiri, K.W. NIH Image to Image J: 25 Years of Image Analysis. *Nat. Methods* **2012**, *9*, 671–675. [CrossRef]
25. Fagerholt, E. ECorrLab. Available online: https://folk.ntnu.no/egilf/ecorr/doc/acquisition/ecorrlabv4/ecorrlabv4.html (accessed on 3 July 2024).
26. Blaber, J.; Adair, B.; Antoniou, A. Ncorr: Open-Source 2D Digital Image Correlation Matlab Software. *Exp. Mech.* **2015**, *55*, 1105–1122. [CrossRef]
27. *ANSYS® Academic Research Mechanical*, ANSYS 2023, Release 19.2; ANSYS Inc.: Canonsburg, PA, USA, 2018.
28. Thirumump, M.; Kalita, K.; Ramachandran, M.; Ghadai, R. A Numerical Study of SCF Convergence Using ANSYS. *J. Eng. Appl. Sci.* **2015**, *10*, 2233–2238.
29. Zheng, Q.; Mashiwa, N.; Furushima, T. Evaluation of Large Plastic Deformation for Metals by a Non-Contacting Technique Using Digital Image Correlation with Laser Speckles. *Mater. Des.* **2020**, *191*, 108626. [CrossRef]
30. Bartczak, Z.; Galeski, A. Plasticity of Semicrystalline Polymers. *Macromol. Symp.* **2010**, *294*, 67–90. [CrossRef]
31. *API 17J 2014*; American Petroleum Institute Specification for Unbonded Flexible Pipe. American Petroleum Institute: Washington, DC, USA, 2014.
32. Defebvin, J.; Barrau, S.; Stoclet, G.; Rochas, C.; Lefebvre, J.M. In Situ SAXS/WAXS Investigation of the Structural Evolution of Poly(Vinylidene Fluoride) upon Uniaxial Stretching. *Polymer* **2016**, *84*, 148–157. [CrossRef]
33. Castagnet, S.; Gacougnolle, J.L.; Dang, P. Correlation between Macroscopical Viscoelastic Behaviour and Micromechanisms in Strained α Polyvinylidene Fluoride (PVDF). *Mater. Sci. Eng. A* **2000**, *276*, 152–159. [CrossRef]
34. Castagnet, S.; Girault, S.; Gacougnolle, J.L.; Dang, P. Cavitation in Strained Polyvinylidene Fluoride: Mechanical and X-Ray Experimental Studies. *Polymer* **2000**, *41*, 7523–7530. [CrossRef]
35. *ASTM D638-14*; Standard Test Method for Tensile Properties of Plastics. ASTM: West Conshohocken, PA, USA, 2014; pp. 1–9.
36. Lakes, R.S.; Wineman, A. On Poisson's Ratio in Linearly Viscoelastic Solids. *J. Elast.* **2006**, *85*, 45–63. [CrossRef]
37. Goolsby, R.D.; Chatterjee, A.M. Notch Sensitivity and Fractography of Polyolefins. *Polym. Eng. Sci.* **1983**, *23*, 117–124. [CrossRef]
38. Pilkey, W.D.; Pilkey, D.F. *Peterson's Stress Concentration Factors*, 3rd ed.; John Wiley & Sons: New York, NY, USA, 2009; ISBN 9780471032212.
39. Challier, M.; Besson, J.; Laiarinandrasana, L.; Piques, R.; Hochstetter, G.; Pignoc, S. Damage and Fracture of Polyvinylidene Fluoride (PVDF) at 20 °C: Experiments and Modelling. *Eng. Fract. Mech.* **2006**, *73*, 79–90. [CrossRef]
40. Maier, G.A.; Wallner, G.; Lang, R.W.; Fratzl, P. Structural Changes during Plastic Deformation at Crack Tips in PVDF Films: A Scanning X-Ray Scattering Study. *Macromolecules* **2005**, *38*, 6099–6105. [CrossRef]
41. Liu, M. *Experimental Studies of the Mechanical Behavior of a PVA Dual-Crosslink Hydrogel*; Faculty of the Graduate School of Cornell University: Ithaca, NY, USA, 2019.

polymers

MDPI

Article

Stress Analysis and Strength Prediction of Carbon Fiber Composite Laminates with Multiple Holes Using Cohesive Zone Models

Hamzah Alharthi [1] and Mohammed Y. Abdellah [2,3,*]

1 Mechanical Engineering Department, College of Engineering and Architecture, Umm Al-Qura University, Makkah 21955, Saudi Arabia; haharthi@uqu.edu.sa
2 Mechanical Engineering Department, Faculty of Engineering, South Valley University, Qena 83523, Egypt
3 Mechanical Engineering Department, College of Engineering, Alasala University, Dammam 31483, Saudi Arabia
* Correspondence: mohamed_abdalla@eng.svu.edu.eg

Abstract: Composite materials play a crucial role in various industries, including aerospace, automotive, and shipbuilding. These materials differ from traditional metals due to their high specific strength and low weight, which reduce energy consumption in these industries. The damage behavior of such materials, especially when subjected to stress discontinuities such as central holes, differs significantly from materials without holes. This study examines this difference and predicts the damage behavior of carbon fiber composites with multiple holes using a progressive damage model through finite element analysis (FEM). Two holes were positioned along the central axis of symmetry in the longitudinal and transverse directions relative to the load. The presence of additional holes acts as a stress-relief factor, reducing stress by up to 17% when the holes are arranged in the longitudinal direction. A cohesive zone model with two parameters, including constant and linear shapes, was applied to develop a simple analytical model for calculating the nominal strength of multi-hole composite laminates, based on the unnotched plate properties of the material. The results closely match experimental findings. The data also provide design tables that can assist with material selection.

Keywords: cohesive zone; damage; carbon fiber; stress intensity factor

1. Introduction

Composite laminate structures play a key role in various industrial applications, such as the automotive, aerospace, and marine industries. This is because of their advantageous properties, including low specific weight, excellent resistance to corrosion and erosion, and high fracture toughness under load [1].

The nominal strength of composite structures with open holes is affected by the size effect phenomenon [1], which refers to a reduction in strength as the structure's geometry is scaled [2]. The nominal strength of composite laminates is influenced by factors such as hole size [3], hole shape [4], and orientation [5,6]. Mohammed [7] simulated the nominal strength of composite laminates with central holes using cohesive laws with two parameters. The study found that the shape of the holes plays a significant role in strength performance, with the constant cohesive law having a greater effect on the fracture toughness (G_{IC}) compared to the linear and exponential cohesive laws.

The performance of composite laminates was enhanced by introducing additional holes in the longitudinal direction [8]. This improvement is attributed to a reduction in stress concentration around the central holes when two holes are present. However, the failure modes in laminates with two holes include delamination and matrix failure, which differ from the behavior observed with a single hole.

A 2D finite element analysis (FEA) was conducted on a woven composite plate with multiple holes. The study revealed that, in non-staggered configurations, the maximum stress at the outer holes causes failure of the net cross-sectional area, whereas staggered configurations depend on the smallest net cross-sectional area. The numerical modeling showed good agreement with the experimental results, accurately predicting crack locations and stress behavior [9]. The arrangement of the holes significantly affects the strength of composite plates. Plates with double-row configurations have smaller net tensile areas and lower strength compared to those with single-row holes [10]. Progressive damage analysis was used to study the failure and damage of laminates [11–13]. The progressive damage model was applied based on the Hashin criteria to analyze the failure of notched laminates. This model successfully evaluated the damage and predicted the ultimate tensile strength for different lay-ups by combining it with nonlinear FEA for laminate systems with circular holes under tension [14]. The damage and failure of composite laminates under stress have been studied using various models, such as continuum damage mechanics [15,16] and FEA combined with the extended finite element method, which has proven to be a simple and effective technique for practical applications [17–19].

In 2021, the nominal strength of an open-hole specimen was predicted using a non-destructive experimental method based on the natural vibration of composite laminates [20]. The core idea of this method was to utilize natural vibrations induced by a lobster blow in an extracted model to calculate the mechanical and fracture properties of composite laminates. Developing a model that can quickly predict the behavior of specimens with holes is crucial for designers to select appropriate materials for specific applications.

Most of the previous studies have focused on predicting tensile strength, whereas a reliable model for predicting failure strength in compression is scarcely available. Therefore, this work was conducted to achieve three main objectives: (1) to predict the size effect in composite laminates with central loading using the progressive damage model, (2) to predict the nominal strength of composite plates with two holes in the longitudinal and transverse directions, considering different distances between the holes, and (3) to investigate the effect of an additional hole on strength reduction or stress relief.

The structure of this paper is as follows: the Section 1 presents the finite element model, the Section 2 outlines the progressive damage model, the Section 3 explains the cohesive laws, and finally, the results and discussion are provided.

2. Progressive Damage Model

2.1. Hashin Damage Model

There are four different failure modes proposed by Hashin. These modes are as follows: fiber compression, fiber tension, and matrix cracking due to compression or tension [21]. The modes are as follows:

Mode I: Fiber tension

$$F_1 = \left(\frac{\overline{\sigma}_{11}}{\chi^{\mathrm{T}}}\right)^2 + \alpha\left(\frac{\overline{\sigma}_{11}}{S^{\mathrm{T}}}\right)^2, \text{ where } 0 \leq \alpha \geq 1 \tag{1}$$

Mode II: Fiber compression

$$F_1 = \left(\frac{\overline{\sigma}_{11}}{\chi^{\mathrm{C}}}\right)^2 \tag{2}$$

Mode III: Matrix tension

$$F_{III} = \left(\frac{\overline{\sigma}_{22}}{Y^{T}}\right)^{2} + \alpha\left(\frac{\overline{\sigma}_{12}}{S^{T}}\right)^{2} \quad (3)$$

Mode IV: Matrix compression

$$F_{IV} = \left(\frac{\overline{\sigma}_{22}}{2S^{T}}\right)^{2} + \alpha\left[\left(\frac{\overline{Y}_{C}}{2S^{T}}\right)^{2} - 1\right]\left(\frac{\overline{\sigma}_{22}}{2\overline{Y}_{C}}\right) + \left(\frac{\overline{\sigma}_{12}}{S^{L}}\right)^{2} \quad (4)$$

where $\overline{\sigma}_{11}$, $\overline{\sigma}_{22}$, and $\overline{\sigma}_{12}$ are principal stress components at each ply and Y^{T}, χ^{C}, $\overline{Y}_{C}$, S^{L}, and S^{T} are composite laminate strengths. The material's elastic and mechanical properties are listed in Tables 1 and 2, which were considered the most important properties for implementing Hashin's damage criteria in the FEM. This model was not a standalone model as it cannot predict the failure of fiber–matrix interaction accurately [22,23], unless it is combined with other damage evaluation laws or degradation rules [24].

Table 1. Elastic constant of T800/924C carbon fiber/epoxy [25].

Properties	E_1 (GPa)	E_2 (GPa)	E_3 (GPa)	μ_{13}	μ_{12}	μ_{23}	G_{12} (GPa)	G_{13} (GPa)	G_{23} (GPa)
Value	161	9.25	9.25	0.34	0.34	0.34	6	6	56

Table 2. Mechanical properties and fracture toughness of $[(\pm 45/0_2)_3]_{2s}$ carbon fiber composite laminates [6].

Properties	X^T MPa	X^C MPa	Yc MPa	S^L MPa	S^T MPa	$(G_{IC})t$ kJ/m^2	$(G_{IC})c$ kJ/m^2	$(G_{IC})tM$ kJ/m^2	$(G_{IC})cM$ kJ/m^2
Value	1615	63	225	109 [26]	109	31	31	0.575	50.575

2.2. Damage Evaluation Criteria

Every failure involves crack initiation and crack growth, both of which are governed by the principles of damage development. Crack initiation is described by the Hashin model, which assumes a linear response. Once failure starts, the damage initiation criteria are applied according to the damage evolution laws, which also reduce the need for refining mesh sizes [27–29]. In Figure 1, a typical linear tensile separation response is shown, as defined by the four failure modes of Hashin's progressive damage criteria. The slope of the first line represents the start of damage, corresponding to the elastic region. The second slope, which is negative, indicates failure beyond the elastic region, commonly referred to as the softening behavior in polymers.

The variable D can be calculated as follows:

$$D = \frac{\delta_{eq}^{f}\left(\delta_{eq} - \delta_{eq}^{o}\right)}{\delta_{eq}\left(\delta_{eq}^{f} - \delta_{eq}^{o}\right)} \quad (5)$$

where D represents the damage variable factor and δ_{eq}^{o} and δ_{eq}^{f} are the equivalent displacements at the end of initiation and the end of failure, respectively.

A sudden degradation rule proposed by Xiao and Ishikawa [24] was implemented immediately after a failure occurred. This rule was designed to prevent the inner ply from carrying specific loads [30].

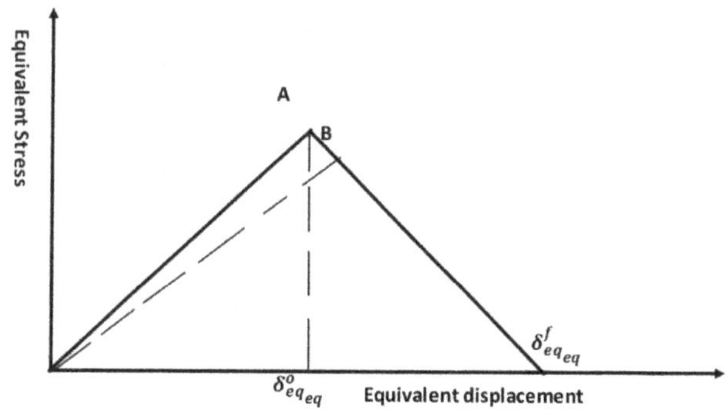

Figure 1. Linear damage law. A: The slope of the first line. B: The second slope.

2.3. Cohesive Zone Model

The cohesive zone model was developed based on cohesive stress at the crack tip and the crack opening displacement (δ). It is commonly referred to as the softening function for quasi-brittle materials, such as tough ceramics, ice, and composite laminates. A specimen fails when the crack opening displacement reaches the critical value (δ_C). The most commonly used cohesive laws are two-parameter cohesive laws, which depend on the surface release energy (G_{IC}) and the stress at the crack tip (σ_C) [1]. The cohesive zone model distributes the stored energy in the material around the crack tip. Examples of the most common cohesive laws, i.e., the linear and constant cohesive laws, are shown in Figure 2a,b.

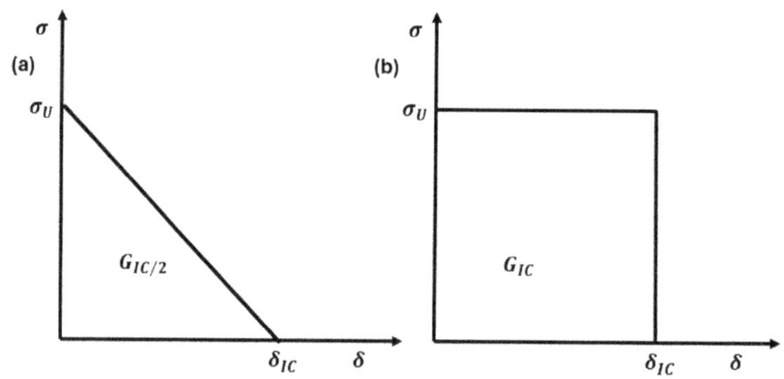

Figure 2. Cohesive zone laws illustrating the relationship between traction and separation: (**a**) linear cohesive law, which shows a linear decrease in traction with increasing separation, and (**b**) constant cohesive law, where the traction remains constant up to a critical separation point.

2.4. Strength Prediction

The problem considered was a composite plate with either a central hole (see Figure 3a), two holes arranged in the transverse direction (Figure 3b), or two holes in the longitudinal direction at a distance (a) apart (Figure 3c). To determine the nominal strength of the composite laminates, the cohesive zone law was applied, using a constant and linear model for specimens with only central holes. Subsequently, a correction factor $f(a/d)$, accounting for the spacing between holes, was calculated. This allowed the nominal strength of composite plates with multiple holes to be calculated. For the complete derivation of the law implementation, please refer to papers [1,7,31].

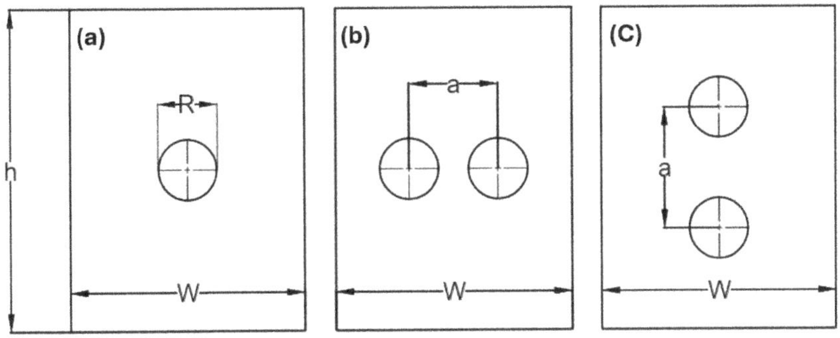

Figure 3. Plate with two holes: (**a**) central hole, (**b**) transversely located hole, and (**c**) longitudinally located hole.

A brief description is provided below.

The generalized cohesive nominal strength equation can be measured as follows [31]:

$$S = \sum_{i=1}^{n} \beta_i \sigma_i \tag{6}$$

$$\beta_i = -\frac{2}{\pi}\left[\sin^{-1}\left(\frac{c_i}{d_s}\right) - \sin^{-1}\left(\frac{b_i}{d_s}\right)\right] f(R/w) \tag{7}$$

where $f(R/W)$ is a correction function and c_i and b_i are displacements at specific points (i) on the crack face found in refs. [1,7,31].

For a constant cohesive law, as shown in Figure 2b, the nominal strength can be calculated using Equation (8). In this case, the stress on the crack face (cohesive stress) is equal to the unnotched tensile strength (σ_u) of the composite laminates.

$$\sigma_N = \frac{\beta_i \sigma_u}{\beta_P} \tag{8}$$

where $\beta_P = 1 - 2R/W$ is the correction factor at the plastic limit. The cohesive stress distribution the over crack face (σ_i) for the linear cohesive law, as shown in Figure 2a, can be calculated using the equation completely derived in [31] and modified with the size effect law in [7].

$$\sigma_i = \sigma_u \left\{1 - \frac{\sigma_u \delta_i}{2\,G_C}\right\} \tag{9}$$

Returning to Equations (6) and (9), the cohesive stress distribution for the linear cohesive law can be calculated using the following Equation (10) [31]:

$$\sigma_i = W_C \left\{\frac{\delta_C}{\sigma_u}\delta_{ij} + f_{ij}\right\}^{-1} I_1 \tag{10}$$

where δ_C represents the critical crack opening and f_{ij} represents the connection function completely calculated in refs. [1,7,31,32].

3. Finite Element Model and Experimental Validation

The composite material domain consisted of a rectangular 3D plate with central holes, as shown in Figure 4a, and another configuration with two holes arranged in the transverse and longitudinal directions, illustrated in Figures 4b and 4c, respectively. The distance between the two holes (a) ranged from 3.5 d to 7.5 d, where d, the hole diameter, was 5 mm according to [6]. The sample domain was modeled as a continuum shell, consisting of 3074 elements for the transverse holes and 2974 elements for the longitudinal holes, as shown in Figures 4b and 4c, respectively. A linear SC8R element type was used with dynamic explicit step loading, along with hourglass control and element deletion criteria. A tabulated amplitude was applied. For the sample with a central hole, as shown in Figure 4a, which was used to predict the size effect and validate the model accuracy, the central hole diameters were varied at 5 mm, 7 mm, 10 mm, and 15 mm. The same mesh area was used for all configurations, with a total of 3346, 3330, 3230, and 2962 elements, respectively. The load was applied to one end of the plate, whereas the opposite end was fixed, as illustrated in Figure 4d.

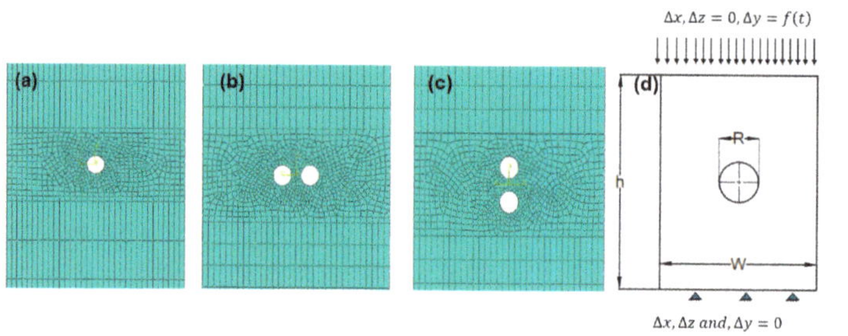

Figure 4. Top view of the 3D model showing: (**a**) the plate with a central open hole, (**b**) two transverse holes, (**c**) two longitudinal holes, and (**d**) the boundary condition domain.

According to Figure 5, which compares three different mesh densities—fine (6692 elements), medium (3346 elements), and coarse (1512 elements)—and studies by the authors [33,34], it was confirmed that the model is independent of mesh refinement. This is because elastic materials exhibit a continuous and uniform stress–strain behavior, ensuring convergence even with coarser meshes. Additionally, the linear stress–strain relationship in elastic materials reduces sensitivity to mesh density.

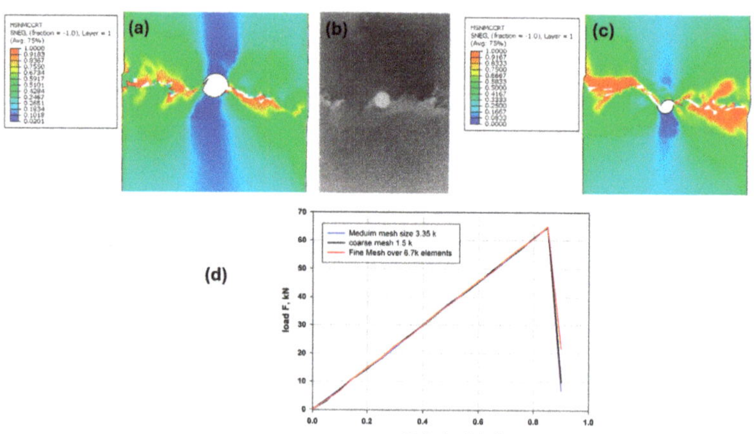

Figure 5. Mesh convergence study with elements size: (**a**) present model with 3k elements, (**b**) experimental [6], (**c**) 6.7k elements, and (**d**) varying mesh size load-carrying capacities.

The primary objective of this study was to evaluate the bearing capacity of composite laminates rather than the mechanisms of delamination behavior. Although finer meshes can capture more detailed crack paths and ply failures, the chosen mesh density was sufficient to achieve accurate results for global bearing capacity, which was the main focus of this investigation. Figure 5a shows the damage contour for the medium mesh compared with the failure mode observed in an experimental study [6] (Figure 5b). In contrast, Figure 5c presents the results for a finer mesh with more than 6000 elements. The differences between the two have minimal impact on accuracy, whereas the medium mesh significantly reduces computational time and cost. This can be attributed to the implementation of the damage evaluation criteria in the model, which ensures reliable results even with moderate mesh densities (see Figure 5d).

The mechanical properties were extracted from composite laminates of T800/924C carbon/epoxy with a stacking sequence $[(\pm 45/0_2)_3]_s$ of 24 layers. It was cut into a plate with a thickness of 3.1 mm, a width of 50 mm, a total length of 245 mm, and a gauge length of 145 mm [6]. The holes should be arranged along the axis of symmetry of the plate, either in the longitudinal or transverse direction, ensuring that $R \leq a \leq w$, as described in [6].

4. Results and Discussion

4.1. Finite Element Results and Cohesive Law Models

Figure 6 shows the predicted nominal strength of the open-hole composite specimens with varying aspect ratios (θ_W). It is clearly observed that strength decreases as the aspect ratio increases, which is known as the size effect or scaling effect [31]. The main objective of simulating the central hole was to validate the Hashin Damage Finite Element Model (HDFEM) with experimental data and the two-parameter cohesive laws, including the constant cohesive law described in Equation (8). The matrix cracking contour and fracture surface predicted by Hashin were similar to those observed in the experimental study by Soutis and Fleck [31], as shown in Figure 7. The maximum stress concentration was clearly observed around the holes, with a crack length proportional to R. The prediction accuracy was over 95% on average, as seen in Figure 6.

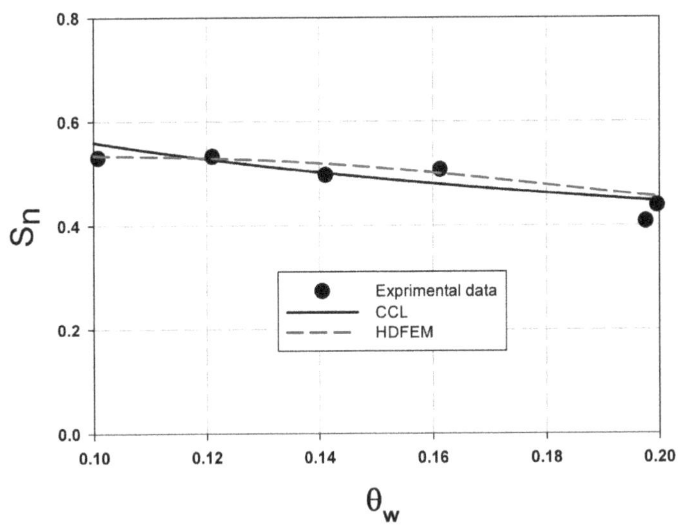

Figure 6. HDFEM validation with the experimental results of Soutis and Fleck [31] for composite laminates with varying hole diameters (varying aspect ratios).

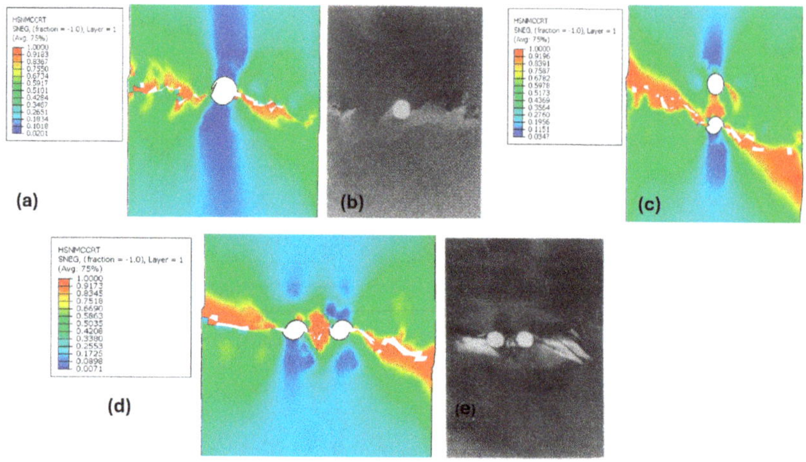

Figure 7. Damage predicted using HDFEM for (**a**) central holes, (**b**) experimental data [6], (**c**) two longitudinally aligned holes, (**d**) two transversely aligned holes, and (**e**) experimental data [6].

Table 3 and Figure 8 show the predicted nominal stresses compared to the experimental data for composite laminates with two aligned transverse holes. The model was found to be very close to the experimental data from Soutis et al. [6]. The percentage error was low (4.72% for the sample with 1.5d spacing between holes), and the smallest error was 0.88% for the case with 2.5 d spacing. The average percentage error was 2.7%, indicating good accuracy and precision in model validation. A fitting linear regression was proposed, which provided resonance data that can be used to determine the correction function $f(a/d)$ in Equations (11)–(14). The same trend was observed for two longitudinally aligned holes, as shown in Table 4 and Figure 9. The maximum percentage error was 5.65% for the case with a larger spacing ratio of 3, whereas the minimum percentage error was 0.42% and the average percentage error was 3.18%. A predictive model is crucial as it helps to make informed decisions, leading to cost savings, higher efficiency, or better results, depending on the application. The percentage error for transversely aligned holes was lower than for longitudinally aligned holes due to more uniform stress distribution, simpler damage mechanisms, and a better match between the assumptions of the damage model and the actual material behavior. In contrast, the more anisotropic stress concentrations and complex damage mechanisms associated with longitudinally aligned holes make the model's predictions less accurate, resulting in a higher percentage error [35,36].

Table 3. HDFEM validation for the specimen with two transverse holes, compared with the experimental work of Soutis et al. [6].

a/d	Nominal Stress σ_r, MPa [6]	HDFEM	% Error
1.5	358	375	4.72
2	413	400	−3.23
2.5	427	423	−0.88
3	465	450	−3.12
3.5	470	463	−1.54

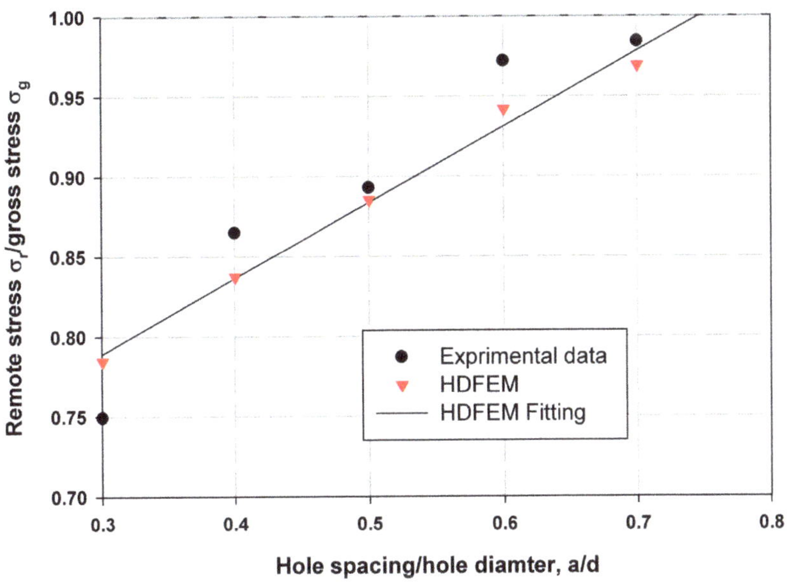

Figure 8. HDM validation with the experimental work of Soutis et al. [6]. for composite laminates with two transverse holes.

Table 4. HDFEM validation for the specimen with two longitudinal holes, compared with the experimental work of Soutis et al. [6].

a/d	Nominal Stress σ_r, MPa [6]	HDFEM	% Error
1.5	516	495	−4.07
2	494.5	480	−2.93
2.5	473	475	0.42
3	451.5	477	5.65
3.5	438.6	451	2.83

The failure modes were simulated through the holes, with the Mises stress shown as a red-colored stress concentration region (see Figure 7). Figure 10 illustrates the effects of existing holes on stress distribution and nominal strength. It was clear that adding additional holes in the longitudinal direction of the central holes reduced the overall strength of the plate. The percentage of strength reduction varied with the distance between the holes. However, for transverse holes, the trend was the opposite, as increasing the distance between the holes resulted in an increase in strength.

This is because, with additional holes in the longitudinal direction near the central holes, the stress concentration factor becomes less complex, and the structure is strengthened by an increase in load-bearing capacity, particularly with closer spacing. On the other hand, with transverse holes, a greater distance between the holes reduces stress interaction and concentration, allowing for a more uniform stress distribution, which can potentially increase the strength of the plate.

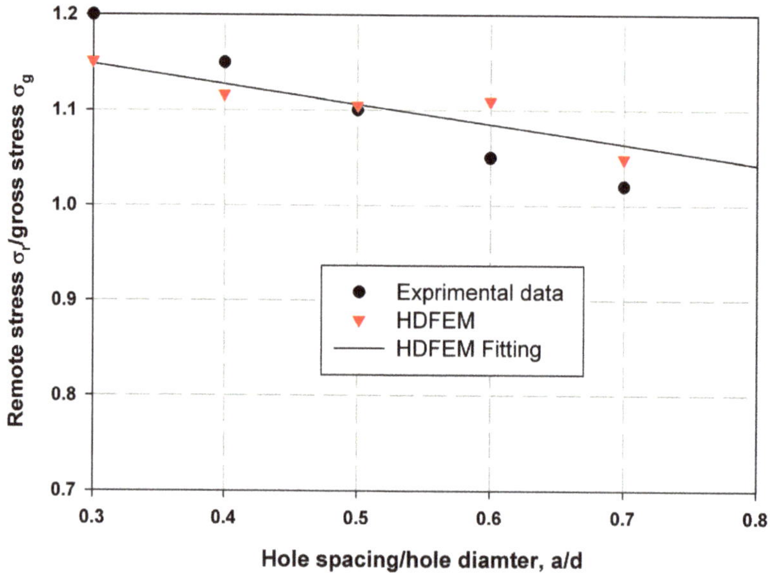

Figure 9. HDM validation with the experimental work of Soutis et al. [6]. for composite laminates with two longitudinal holes.

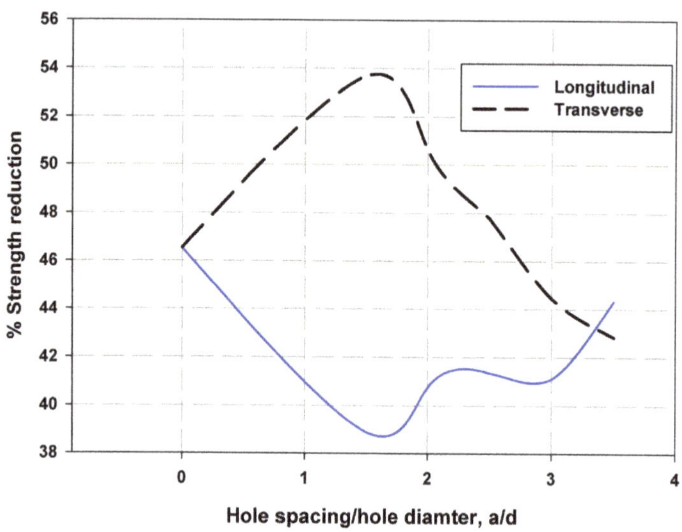

Figure 10. Stress reduction influenced by the insertion of an extra hole.

4.2. Nominal Strength of the Plate with Multiple Holes

The nominal strength of the composite plate with two open holes subjected to compression load can be predicted using the correction function $f(a/d)$, which was obtained by HDFEM fitting through linear regression, as previously discussed.

Constant cohesive law (CCL):

For the constant cohesive law, Equation (8) can be multiplied by the correction factor obtained from curve fitting of the data in Figures 8 and 9, as follows:

Transverse direction:

$$\sigma_N = \frac{\beta_i \sigma_u}{\beta_P} f\left(\frac{a}{d}\right) \tag{11}$$

where the correction function is given by:

$$f\left(\frac{a}{d}\right) = \left(0.47\left(\frac{a}{d}\right) + 0.64\right) \tag{12}$$

Longitudinal direction:

$$\sigma_N = \frac{\beta_i \sigma_u}{\beta_P} \overline{f}\left(\frac{a}{d}\right) \tag{13}$$

where the correction function is calculated as:

$$\overline{f}\left(\frac{a}{d}\right) = \left(-0.21\left(\frac{a}{d}\right) + 1.21\right) \tag{14}$$

Linear cohesive law (LCL):

For the linear cohesive law, Equation (6) can be multiplied by the correction factor obtained from curve fitting of the data in Figures 8 and 9, as follows:

Transverse direction:

$$S = \sum_{i=1}^{n} \beta_i \sigma_i f\left(\frac{a}{d}\right) \tag{15}$$

Longitudinal direction:

$$S = \sum_{i=1}^{n} \beta_i \sigma_i f\left(\frac{a}{d}\right) \tag{16}$$

Figures 11 and 12 show the predicted nominal strength for two transversely oriented holes using the proposed models based on the following cohesive laws: LCL and CCL. CCL was found to provide more accurate data and predictions than LCL for both transversely and longitudinally oriented holes. The average error for CCL was 2.5%, whereas for LCL, it was 9% when the two holes were transversely aligned, as shown in Figure 10.

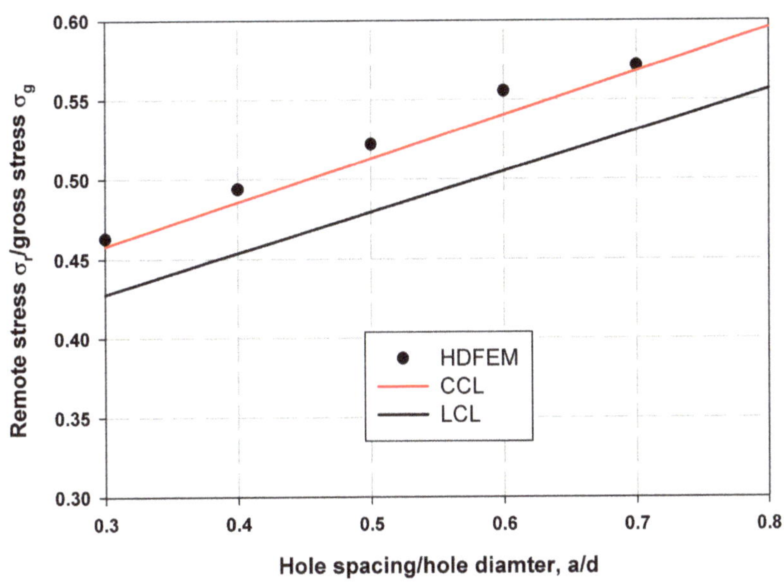

Figure 11. Model validation comparing the predicted and HDFEM results for a structure with two transverse holes.

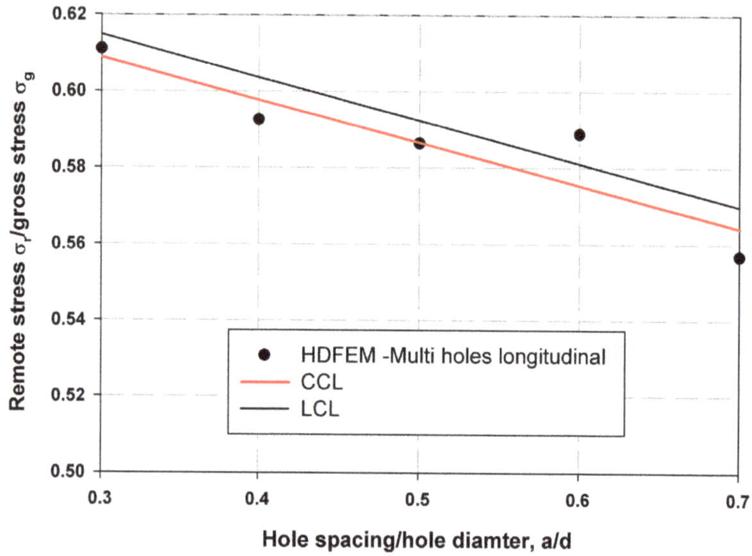

Figure 12. Model validation comparing the predicted and HDFEM results for a structure with two longitudinal holes.

However, in the case of two longitudinally aligned holes, as observed in Figure 11, the percentage error for LCL was larger than that for CCL. Specifically, the error was 1.2% for LCL and 0.5% for CCL. The difference between LCL and CCL can be attributed to several factors. The first reason is the surface release energy, where the linear cohesive law assumes a varying pressure–separation relationship, reflecting the increasing resistance of the material as the crack tip approaches the failure point. In contrast, the constant cohesive law assumes a constant compressive force during crack propagation, resulting in a simpler and typically higher nominal strength prediction. This difference in crack growth behavior explains the observed differences in the predicted strengths [37]. Additionally, the symmetrical holes, which were oriented transverse to the grain direction, led to a more uniform stress distribution in the specimens.

5. Conclusions

The Hashin Damage Finite Element Model (HDFEM), also known as the progressive damage model, predicts the size effect on the compressive response of a composite plate with scaled geometry. Additionally, the progressive damage model helps to determine the correction factor for predicting the nominal strength of composite plates with multiple aligned holes in both transverse and longitudinal directions. The model provides an average prediction accuracy of 2.5%.

The proposed model, based on the linear and constant cohesive laws with two parameters, provides highly accurate results, with a minimum average error of 0.5% in the case of the linear cohesive law for two holes aligned in the longitudinal direction. The presence of additional holes in the longitudinal direction decreases the overall strength ratio. It acts as a stress relief factor, reducing stress concentration around the central hole region. Specifically, the percentage reduction in strength increases to 17% in the case of two holes arranged in the longitudinal direction.

The model provides valuable tables and diagrams that are important for design considerations and material selection.

Author Contributions: Conceptualization, M.Y.A. and H.A.; methodology, M.Y.A. and H.A.; software, M.Y.A. and H.A.; validation, M.Y.A.; formal analysis, M.Y.A. and H.A.; investigation, M.Y.A.; resources, H.A.; writing—original draft preparation, M.Y.A.; writing—review and editing, H.A.; supervision, M.Y.A.; project administration, M.Y.A.; funding acquisition, H.A. All authors have read and agreed to the published version of the manuscript.

Funding: This research received no external funding.

Institutional Review Board Statement: Not applicable.

Data Availability Statement: Data are contained within the article.

Conflicts of Interest: The authors declare no conflicts of interest.

List of Nomenclature

Symbols	Definitions
θ_W	Aspect ratio equals R/w
σ_C	Cohesive stress
σ_i	Component of stress for the *i*-th element on the crack face
Y^T, χ^C, $\overline{Y}_C$, $S^L$and S^T	Composite laminate strengths
f_{ij}	Connection function calculated for the element i, j
β_P	Correction factor at the plastic limit
$f(a/d)$	Correction factor for spacing between holes
δ_C	Critical opening displacement
D	Damage variable factor
c_i, b_i	Displacements at specific points (i) on the crack face, which are equal to $c_i = (R + \iota) - \left(\frac{i-1}{n}\right)$ $b_i = (R + \iota) - \left(\frac{i}{n}\iota\right)$
E_1, E_2, E_3, μ_{12}, μ_{13}, μ_{23}	Elastic constants for T800/924C carbon fiber/epoxy
δ^o_{eq}, and δ^f_{eq}	Equivalent displacement at the initiation and failure stages
F_1. F_{11}, F_{111}, and F_{1V}	Functions represent fiber tension, fiber compression, matrix tension, and matrix compression
R	Hole radius
δ_{ij}	Kronecker delta in continuum mechanics
d_s	Length from hole centers to the crack tip, which is equal to $d_s = l + R$
l	Length of micro-buckled region
S_n	Nominal strength
δ	Opening displacement
w	Plate width
$\overline{\sigma}_{11}$, $\overline{\sigma}_{22}$and $\overline{\sigma}_{12}$	Principal stress components at each ply
σ_r	Remote strength or nominal strength (Sn)
β_i	Shape factor
G_{IC}	Surface release energy
σ_u	Un-notch composite strength
$f(R/w)$	Width and hole correction function

References

1. Hassan, M.K.; Mohammed, Y.; Salem, T.; Hashem, A. Prediction of nominal strength of composite structure open hole specimen through cohesive laws. *Int. J. Mech. Mech. Eng.* **2012**, *12*, 1–9.
2. Bažant, Z.P. Size effect. *Int. J. Solids Struct.* **2000**, *37*, 69–80. [CrossRef]
3. Lee, J.; Soutis, C. Measuring the notched compressive strength of composite laminates: Specimen size effects. *Compos. Sci. Technol.* **2008**, *68*, 2359–2366. [CrossRef]
4. Daynes, S.; Das, R. Ultimate strength prediction of composite laminates containing straight-sided holes, scarfed holes, and bonded repairs. *Compos. Struct.* **2024**, *348*, 118512. [CrossRef]

5. Özaslan, E.; Güler, M.A.; Yetgin, A.; Acar, B. Stress analysis and strength prediction of composite laminates with two interacting holes. *Compos. Struct.* **2019**, *221*, 110869. [CrossRef]
6. Soutis, C.; Fleck, N.; Curtis, P. Hole-hole interaction in carbon fibre/epoxy laminates under uniaxial compression *Composites* **1991**, *22*, 31–38. [CrossRef]
7. Mohammed, Y.; Hassan, M.K.; Abu El-Ainin, H.; Hashem, A.M. Size effect analysis of open-hole glass fiber composite laminate using two-parameter cohesive laws. *Acta Mech.* **2015**, *226*, 1027–1044. [CrossRef]
8. Khan, R.M.A.; Shafighfard, T.; Ali, H.Q.; Mieloszyk, M.; Yildiz, M. Strength prediction and experimental damage investigations of plain woven CFRPs with interacting holes using multi-instrument measurements. *Polym. Compos.* **2023**, *44*, 3594–3609. [CrossRef]
9. Supar, K.; Ahmad, H. Stress distribution study on multi-holes configurations in woven fabric kenaf composite plates. In Proceedings of the IOP Conference Series: Materials Science and Engineering, Johor Bahru, Malaysia, 28–29 August 2017; IOP Publishing: Bristol, UK, 2017; p. 012005.
10. Ahmad, H. Notched Strength of Double-Row Multi-Holes Woven Fabric Kenaf Composite Plates with Different Plate Orientations. *Adv. Res. Nat. Fibers* **2019**, *1*.
11. Zhang, D.; Ye, J.; Lam, D. Ply cracking and stiffness degradation in cross-ply laminates under biaxial extension, bending and thermal loading. *Compos. Struct.* **2006**, *75*, 121–131. [CrossRef]
12. Bouazza, M.; Tounsi, A.; Benzair, A.; Adda-Bedia, E. Effect of transverse cracking on stiffness reduction of hygrothermal aged cross-ply laminates. *Mater. Des.* **2007**, *28*, 1116–1123. [CrossRef]
13. Spottswood, S.M.; Palazotto, A.N. Progressive failure analysis of a composite shell. *Compos. Struct.* **2001**, *53*, 117–131. [CrossRef]
14. Chang, F.-K.; Chang, K.-Y. A progressive damage model for laminated composites containing stress concentrations. *J. Compos. Mater.* **1987**, *21*, 834–855. [CrossRef]
15. Maimí, P.; Camanho, P.P.; Mayugo, J.; Dávila, C. A continuum damage model for composite laminates: Part I–Constitutive model. *Mech. Mater.* **2007**, *39*, 897–908. [CrossRef]
16. Camanho, P.P.; Maimí, P.; Dávila, C. Prediction of size effects in notched laminates using continuum damage mechanics. *Compos. Sci. Technol.* **2007**, *67*, 2715–2727. [CrossRef]
17. Rybicki, E.F.; Kanninen, M.F. A finite element calculation of stress intensity factors by a modified crack closure integral. *Eng. Fract. Mech.* **1977**, *9*, 931–938. [CrossRef]
18. Krueger, R. Virtual crack closure technique: History, approach, and applications. *Appl. Mech. Rev.* **2004**, *57*, 109–143. [CrossRef]
19. Abdellah, M.Y.; Alsoufi, M.S.; Hassan, M.K.; Ghulman, H.A.; Mohamed, A.F. Extended finite element numerical analysis of scale effect in notched glass fiber reinforced epoxy composite. *Arch. Mech. Eng.* **2015**, *62*, 217–236. [CrossRef]
20. Abdellah, M.Y.; Hassan, M.K.; Mohamed, A.F.; Khalil, K.A. A novel and highly effective natural vibration modal analysis to predict nominal strength of open hole glass fiber reinforced polymer composites structure. *Polymers* **2021**, *13*, 1251. [CrossRef] [PubMed]
21. Hashin, Z. Failure criteria for unidirectional fiber composites. *J. Appl. Mech.* **1980**, *47*, 329–334. [CrossRef]
22. Dandekar, C.R.; Shin, Y.C. Modeling of machining of composite materials: A review. *Int. J. Mach. Tools Manuf.* **2012**, *57*, 102–121. [CrossRef]
23. Abdellah, M.Y.; Hassan, M.K. Numerical Analysis of Open Hole Specimen Glass Fiber Reinforced Polymer. *Nonlinear Eng.* **2014**, *3*, 141–147. [CrossRef]
24. Xiao, Y.; Ishikawa, T. Bearing strength and failure behavior of bolted composite joints (part II: Modeling and simulation). *Compos. Sci. Technol.* **2005**, *65*, 1032–1043. [CrossRef]
25. Soutis, C.; Fleck, N. Static compression failure of carbon fibre T800/924C composite plate with a single hole. *J. Compos. Mater.* **1990**, *24*, 536–558. [CrossRef]
26. Soutis, C.; Lee, J. The effect of specimen size on the compressive strength of carbon fibre-epoxy laminates. In *Recent Advances in Experimental Mechanics*; Springer: Dordrecht, The Netherlands, 2002; pp. 153–162.
27. Abdellah, M.Y.; Gelany, A.F.; Mohamed, A.F.; Khoshaim, A.B. Protection of limestone Coated with Different Polymeric Materials. *Am. J. Mech. Eng.* **2017**, *5*, 51–57. [CrossRef]
28. Abdellah, M.Y. Ductile Fracture and S–N Curve Simulation of a 7075-T6 Aluminum Alloy under Static and Constant Low-Cycle Fatigue. *J. Fail. Anal. Prev.* **2021**, *21*, 1476–1488. [CrossRef]
29. Abdellah, M.Y. Essential Work of Fracture Assessment for Thin Aluminium Strips Using Finite Element Analysis. *Eng. Fract. Mech.* **2017**, *179*, 190–202. [CrossRef]
30. Kermanidis, T.; Labeas, G.; Tserpes, K.; Pantelakis, S. Finite element modeling of damage accumulation in bolted composite joints under incremental tensile loading. In Proceedings of the Third ECCOMAS Congress, Barcelona, Spain, 11–14 September 2000.
31. Soutis, C.; Fleck, N.; Smith, P. Failure prediction technique for compression loaded carbon fibre-epoxy laminate with open holes. *J. Compos. Mater.* **1991**, *25*, 1476–1498. [CrossRef]
32. Soutis, C.; Curtis, P. A method for predicting the fracture toughness of CFRP laminates failing by fibre microbuckling. *Compos. Part A Appl. Sci. Manuf.* **2000**, *31*, 733–740. [CrossRef]

33. Alssayegh, A.; Abdellah, M.Y.; Hassan, M.K.; Azam, S.; Melaibari, A.; Khashaba, U. Optimizing High Cycle Fatigue Predictions in Notched Al 7075-T6: An Analytical Approach to Rotating Bending Behavior. *Results Eng.* **2024**, *25*, 103623. [CrossRef]
34. Abdellah, M.Y.; Alharthi, H. Fracture Toughness and Fatigue Crack Growth Analyses on a Biomedical Ti-27Nb Alloy under Constant Amplitude Loading Using Extended Finite Element Modelling. *Materials* **2023**, *16*, 4467. [CrossRef] [PubMed]
35. Hashin, Z.; Shtrikman, S. A variational approach to the theory of the elastic behaviour of multiphase materials. *J. Mech. Phys. Solids* **1963**, *11*, 127–140. [CrossRef]
36. Christensen, R.M. *Mechanics of Composite Materials*; Courier Corporation: Chelmsford, MA, USA, 2005.
37. Xu, X.-P.; Needleman, A. Numerical simulations of fast crack growth in brittle solids. *J. Mech. Phys. Solids* **1994**, *42*, 1397–1434. [CrossRef]

polymers

MDPI

Article

Advanced Prediction and Analysis of Delamination Failure in Graphite-Reinforced Epoxy Composites Using VCCT-Based Finite Element Modelling Techniques

Ahmed F. Mohamed [1], Mohammed Y. Abdellah [2,3,*], Mohamed K. Hassan [4] and Ahmed H. Backar [4,5]

1 Industrial Engineering Department, College of Engineering and Architecture, Umm Al-Qura University, P.O. Box 5555, Makkah 21955, Saudi Arabia; afmohamed@uqu.edu.sa
2 Mechanical Engineering Department, Faculty of Engineering, South Valley University, Qena 83523, Egypt
3 Mechanical Engineering Department, College of Engineering, Alasala Colleges, Dammam 31483, Saudi Arabia
4 Mechanical Engineering Department, College of Engineering and Architecture, Umm Al-Qura University, P.O. Box 5555, Makkah 21955, Saudi Arabia; mkibrahiem@uqu.edu.sa
5 Production Engineering Department, Faculty of Engineering, Alexandria University, Alexandria 21544, Egypt; ahmed.backar@alexu.edu.eg
* Correspondence: mohamed_abdalla@eng.svu.edu.eg

Abstract: The applications of graphite-reinforced composite laminates have gained significant importance since the last century and remain a highly attractive field due to their widespread and versatile applications. Among the various failure modes, delamination—defined as the separation of layers within the composite structure—stands out as the most common and critical type of failure in these materials. In this study, the mode I interlaminar fracture energy was predicted using the virtual crack closure technique (VCCT) integrated with a finite element model (FEM), applied to a double cantilever beam (DCB) specimen. Additionally, a straightforward analytical model was developed to calculate the critical fracture energy in mode I. The analytical model used the material strength and stiffness. The results demonstrated strong agreement with experimental data, with a margin of error as low as 5%, highlighting the accuracy and reliability of the proposed methods.

Keywords: virtual crack closure technique; surface release energy; graphite reinforcement; double cantilever beam

1. Introduction

Composite laminates are strong challenges to many traditional metals like steel and aluminium due to their excellent specific strength, light weight, and high resistance to corrosion and erosion. Delamination, which is defined as the separation of layers, is considered the primary and most serious failure mode in these materials. Therefore, significant effort must be assigned to understanding and predicting this phenomenon. In recent years, various methods and models have been proposed to predict delamination growth [1,2] and transverse cracking [3–5].

Many studies and approaches with different names have considered the problem of delamination, such as the cohesive fracture model [6], cohesive layer [7–9], and interfacial decohesion [8]. The virtual crack closure technique (VCCT) [10] is widely recognized as the most effective tool to approximate the surface release energy G. This method requires determination of the nodal forces at the delamination tip, the displacement openings directly behind it, and the virtual area in front of the crack tip. To calculate the release energy G for the mixed-mode in a general case, the nodal forces and the displacement openings must be transformed into an instantaneous local coordinate system aligned with the delamination tip, which is only possible after determining the normal direction over delamination path [11].

The delamination usually gets measured using a double cantilever beam (DCB) specimen [12,13]. The finite element method was widely used for the simulation of delamination failure using surface-based cohesive [11], cohesive element [14,15], or even virtual crack closure techniques [16,17]. Delamination failure is important both when the laminated composite is in service and through machining or cutting like drilling [13]. In a model by Li [18], delamination and transverse crack growth in laminated composite panels and shells was predicted. It considered interlaminar and intralaminar damage mechanisms using fracture mechanics and cohesive zone modelling. The approach considered material properties, loading conditions, and geometric factors that influence crack propagation. Numerical simulations validated the model against experimental data and proved its accuracy. The results improved the understanding of composite failure and helped in the design of more durable structures. Whitcomb [19] investigated delamination growth in composite plates due to local buckling using a 3D nonlinear finite element model and a strain energy release rate-based fracture criterion. Chen [20] examined elastic buckling and post-buckling behaviors in an axially loaded beam plate with transverse delamination, employing shear deformation theory and a Griffith-type fracture criterion, confirming the reduction in critical buckling and ultimate loads due to shear deformation. Hitchings et al. [21] introduced a finite element technique for modelling arbitrarily shaped delamination growth in laminated composites. They combined linear elastic fracture mechanics (LEFM) and a strain energy release rate-based criterion to be applied along the delamination tip. Aymerich et al. [22] employed the virtual internal bond model to predict delamination initiation and growth in unidirectional laminated composites, including DCB, end notch flexure (ENF), and mixed-mode bending (MMB) tests. The staking sequences layup geometry was used to calculate the fiber tension surface release energy [23], and a simple analytical model was extracted, which had a high degree of accuracy. Additionally, researchers attempted to calculate the fracture properties using non-traditional testing operations for composite laminates using natural frequencies of strikes on a DCB [24]. Meer et al. [25] proposed a method for progressive delamination modelling using a level set field to implicitly define the crack front, introducing weak discontinuities to capture transitions between cracked and uncracked regions, and applied an explicit energy-based relation where crack growth was determined by the configurational force derived from other mathematical models.

As mentioned before, there is a great need for a deeper understanding of the failure modes of composite laminates, so the present study deals with three main objectives: (1) development of a Virtual Crack Closure technique in combination with FEM to simulate delamination failure, (2) analytical prediction of the resistance curve using the VCCT and the presented equations, and (3) prediction of the failure load after reaching the peak load, where only the material strength and the beam geometry are known.

The work was structured as follows: In Section 2, the VCCT was outlined, then the analytical model was extracted in Section 3, followed by Section 4, in which the corresponding finite element model was extracted. In Section 5, the results were correlated and then the conclusions and adjustments were outlined.

2. Virtual Crack Closure Technique (VCCT)

The basic idea of the (VCCT), a popular technique for calculating the energy release rate and modelling crack propagation in fracture mechanics, is depicted in the drawing in Figure 1. It specifically shows the separation at a virtually opened fracture tip and the nodal response forces, which are crucial components in assessing the stress intensity parameters. In the situation of VCCT, the technique assumes that the energy required to close a virtually opened crack equals the energy needed to propagate the crack further. This assumption is modelled by computing the forces acting on the crack faces (nodal reaction forces) and the corresponding displacements or separations at these nodes. The interaction between the nodal forces and separations provides a direct estimation of the strain energy release rate, as proposed by Rybicki and Kanninen in their work on VCCT [10]. The relative displacements between neighboring nodes at the fracture tip brought on by external loading are represented by the crack face separation, which was depicted in Figure 1. Finding the crack-opening mode (Mode I), shear mode (Mode II), or tearing mode (Mode III), which together characterize the fracture mechanics behavior of the structure being studied, requires these divisions. In a similar manner, the internal stresses that prevent the crack faces from virtually separating are represented by the nodal response forces, which balance the external loading conditions. The VCCT framework, which combines these parameters, offers a computationally efficient and reliable method for predicting the onset and propagation of cracks in a variety of materials and structures. For applications in energy systems, mechanical engineering, and aerospace where the integrity of materials under load is crucial, this makes it especially helpful [26].

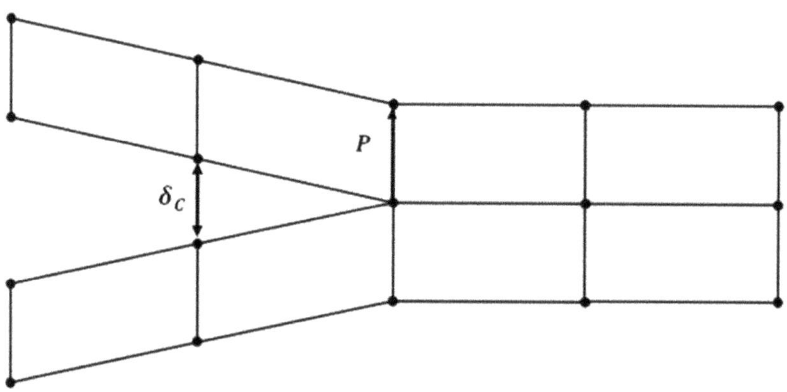

Figure 1. Nodal reaction force (P) and critical separation δ_C, used in Equation (1) of crack tip [27].

2.1. Damage Evaluation Criteria

Delamination failure involves crack initiation and crack growth. These two mechanisms are governed by the principles of damage development. When the load reaches a critical value or the maximum peak load, failure occurs. This initiation criterion was established by linear evaluation laws. These evaluation laws reduce the need for refining mesh sizes [28–30]. A typical linear tensile separation response is shown in Figure 2. The onset of damage is represented by the initial slope, followed by softening, characterized by the negative linear portion of the curve, which indicates failure beyond the elastic region.

The variable D can be calculated as follows:

$$D = \frac{\delta_{eq}^{f}\left(\delta_{eq} - \delta_{eq}^{o}\right)}{\delta_{eq}\left(\delta_{eq}^{f} - \delta_{eq}^{o}\right)} \quad (1)$$

where D damage variable factor, δ_{eq}^{o}, and δ_{eq}^{f} are the equivalent displacements at the end of initiation and the end of failure, respectively.

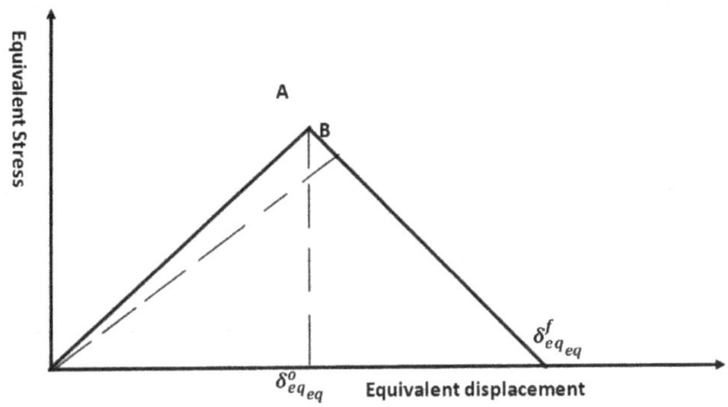

Figure 2. Linear damage law.

2.2. Finite Element Modelling

The DCB test was simulated using two plates with 24.4 mm width, 101.6 mm length, 2.14 mm thickness, and pre-crack length 50.8 mm (see Figure 3a). The model was created as a pair of contact with interaction properties for VCCT as mixed-mode (BK) with maximum tangential stress, and the interfacial mixed model values are listed in Table 1. The surface based cohesive zone (SCZM) uses the same interaction behaviors previously discussed, but the cohesive penalty interfacial stiffness is equal to ($K_o = 4 \times 10^5$ N/mm^3) [31]. The maximum traction separation N = 80 MPa [31]. This method was used to validate the VCCT as previously explained. The implemented initial clearance was 1×10^{-5}. This method was completely described in Abdellah [11]. The linear evaluation law for both methods were linear shapes, as illustrated in Figure 2. A bilinear, four-node, plane stress quadrilateral element (CPS4R) with reduced integration and deletion capability was employed in the finite element analysis. This element offers a good compatibility between computational efficiency and accuracy, and reduced integration element type in numerical stiffness and hypothetical hourglassing improves performance for large-scale simulations. The deletion feature allows for realistic modelling of material failure, crucial for studying crack initiation and propagation. The mesh, detailed in Figure 3b, utilized 1 mm elements with 2244 elements. The load (P) was applied to point 1, and 2, which coupled with the top and bottom surface of the DCB, as shown In Figure 3a. In previous studies [32,33], the mesh convergence was establish to have little effect in the case of measuring the load carrying capacity of composite plate. Additionally, the linear evaluation damage model in elastic materials reduces sensitivity to mesh density.

Table 1. Elastic constant of IM7/8552 carbon fiber/epoxy [34].

Properties	E1	E3 = E2	μ13	μ12	μ23	G12	G13	G23
GPa	152.69	8.703	0.32	0.32	0.45	5.16	5.16	5.16

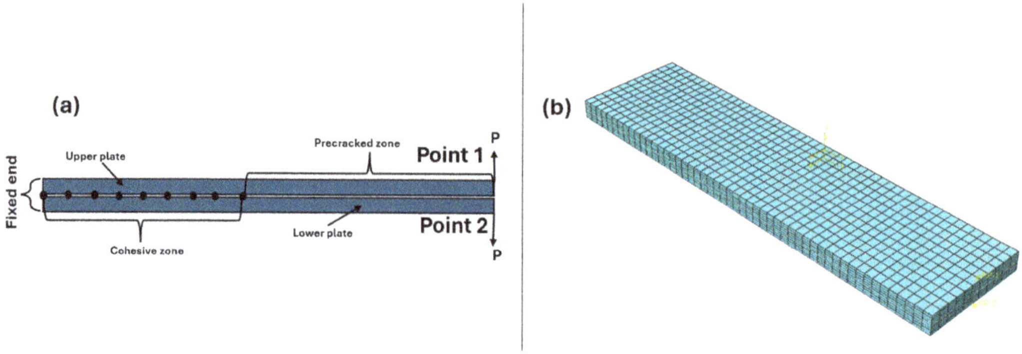

Figure 3. FEM: (**a**) boundary conditions and (**b**) mesh domain.

3. Analytical Model

Delamination

Delamination is the process by which layers in a laminated composite material separate from one another. The load carrying capability of the entire laminate may have been impacted by impact, manufacturing flaws, and cyclic loading. This type of failure typically occurs in applications where composite laminates are widely employed, such as aerospace, automotive, and wind energy applications. Many numerical and experimental studies provide a good understanding of delamination in laminated composite materials [35]. However, further studies are needed to optimize and deepen the understanding of the failure modes in these materials.

The resistance of delamination measured by the interlaminar fracture energy G_{IC} in mode I

$$G_{IC} = \frac{3P\delta_C}{2B(a + |\Delta|)} \tag{2}$$

where P is load, δ_C is the critical displacement (computed using VCCT), B is the specimen width, a is the initial delamination length, and $|\Delta|$ is the absolute correction factor determined form the relationship between compliance and delamination length, which would be predicted using VCCT. Compliance relation according to beam theory can be measured as follows:

$$C^{1/3} = m(a + |\Delta|) \tag{3}$$

where m is the linear regression slope.

In the pre-cracked DCB shown in Figure 4, the arms can be considered like a cantilever beam supported at the right end and loaded at the lift end. The model was based on beam theory [36], and the beam deflection δ (COD) can be calculated using Equation (4) for one DCB arm as follows:

$$\delta = \frac{P \times a^3}{3\,EI} \tag{4}$$

where I is the second moment of inertia, P is the load, and a is the pre-crack length. Additionally, considering the deflection for both DCB arms, and substitution by the value of moment of inertia for rectangular geometry ($I = \frac{Bh^3}{12}$), the following Equation (5) can be used for total deflection:

$$\delta_{Cr} = \frac{4Pa^3}{EBh^3} \tag{5}$$

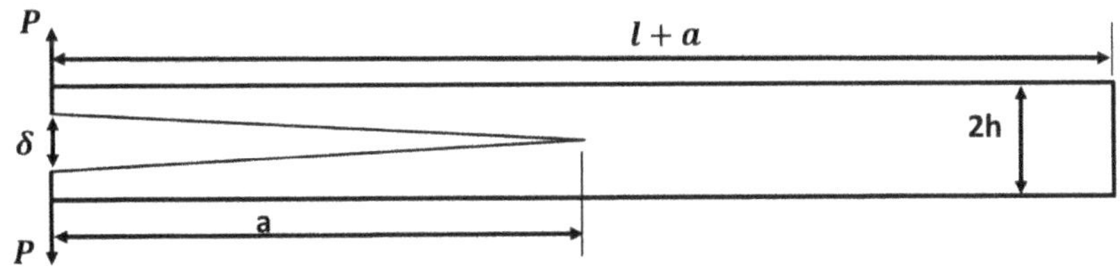

Figure 4. Problem of double cantilever beam.

The compliance of specimen Equation (3), therefore it can be rewritten with respect to applied load and beam stiffness as follows:

$$C = \frac{8a^3}{EBh^3} \tag{6}$$

The surface release energy G_{IC} as stated in Equation (2) related to compliance, therefore it can be rewritten as follows:

$$G_{IC} = \frac{P^2 \partial C}{2\partial A} \tag{7}$$

where ∂A is the crack extension area, and the h is the arm thickness.

By applying the differentiation for compliance Equation (6) and implemented through Equation (6), the surface release energy G_{IC} can be calculated as follows:

$$G_{IC} = \frac{12P^2a^2}{EB^2h^3} \tag{8}$$

The arms of the cantilever beam shown in Figure 4 were subjected to bending moment; therefore, the bending stress can be calculated as follows:

$$M = \frac{\sigma_b\, I}{h} \tag{9}$$

Based on beam theory and linear elastic fracture mechanics fundamentals, the surface release energy can be calculated using the work done and energy stored through material as follows [36]:

$$U_E = 2\int_0^a \frac{M(x)}{2EI} \tag{10}$$

where U_E is the elastic stored energy and $(M(x) = P \times a)$ is the moment function through crack length maximum at $x = a$.

By substituting Equation (10) into Equation (8), considering conditions of the moment at maximum state, the surface release energy can be measured as follows:

$$G_{IC} = \frac{M^2}{BEI} \tag{11}$$

Considering the following state for surface release energy in terms of COD, it can be measured as follows:

$$G_{IC} = 2\delta_{Cr} \times \sigma_b \tag{12}$$

where σ_b is bending stress over beam arms [11] and equal in the case of delamination testing the transverse laminate strength ${Y_{22}}^T$ (see Figure 4), where δ_{Cr} can be measured using VCCT previously explained for each load.

Substituting Equation (9) into Equation (11), and re-editing, the final surface release can be measured as follows:

$$G_{IC} = \frac{{\sigma_b}^2 \times h}{6E} \tag{13}$$

This equation stands alone for testing the surface release energy without the availably of COD; it just needs the specimen geometry and equivalent Young's modulus. To calculate the peak load at which crack initiation takes place. Equating Equation (13) with Equation (8), it can be shown as follows:

$$P = \frac{\sigma_b B \times h^2}{6\sqrt{2} \times a} \quad (14)$$

The model can not be used for multidirectional laminates or for measuring the fiber tension facture toughness because it needs the transverse tensile strength without increasing bridging of fiber, which may change the strength results. Additionally, VCCT assumes that the crack propagates in a predefined direction along the existing mesh. This can be a limitation in composite laminates where crack growth may follow complex, mixed-mode paths due to anisotropic properties. Additionally, VCCT is based on the assumption that a crack already exists and only predicts its propagation. It does not model crack initiation, which is crucial in composite laminates where damage can start as matrix cracking, fiber breakage, or delamination before an actual crack forms. This limitation makes VCCT less effective for predicting the onset of failure in composites [11,17,26].

4. Material

The composite laminates were unidirectional IM7/8552 carbon fiber-reinforced epoxy, and the elastic constant of such materials are listed in Table 1. These mechanical properties would be implemented in the above models. The pre-crack was an insert of 50.8 mm, which was created using 12.5 μm thick Teflon. The Teflon strips were inserted as the final step in the laminate layup. For each specimen, a straight edge was placed on the uncured laminate to mark the exact location of the intralaminar pre-crack. The specimen dimensions were 24.4 mm width, 101.6 mm length, and the specimen thickness nearly 4.34 mm. A complete description of the DCB specimens and experimental procedures would be found in Ref. [34]. The specimens were manufactured using IM7/8552 carbon fiber/epoxy pre-preg tap of 36 ply unidirectional laminates.

5. Results and Discussion

Figure 5 shows the load–displacement relation for DCB measured using VCCT and SCZM compared using the experimental data found in Ref. [34]. It was clear that the prediction was highly close to the experimental data for the two models, while for VCCT it gave a more realistic trend as a zigzag shape corresponding to the loading and unloading action for the ideal test; also, it helps in measuring the compliance of the zigzag slope. Those zigzag and step-like shapes were due to stick–slip crack propagation [37–39], some fiber bridging in the direction of the crack, as the crack does not propagate smoothly, but rather arrests temporarily and then jumps forward when the energy release rate exceeds a critical value. Additionally, the fiber bridging the crack resists the load, therefore the load increases, and when this fiber breaks, the load suddenly decreases, making these zigzags. Additionally, due to the heterogeneity in material properties, as the carbon fiber reinforced polymer (CFRP) was an anisotropic material, they exhibit different toughness and strength through the loading direction. Moreover, it returns to frictional interaction during the fixation of the specimen or through the testing machine and other experimental results [39]. However, the predicted curve gave shorter extension than the experimental ones; this was because real CFRP laminates exhibit matrix plasticity, micro-cracking, and local damage at the crack tip. These parameters were important and serious for increasing COD, while for VCCT these factors or parameters were not included for calculations [40]. Moreover, the test machine compliance and slow crack growth in the experiment give longer and increasing COD; however, this factor were not considered in the case of VCCT [39].

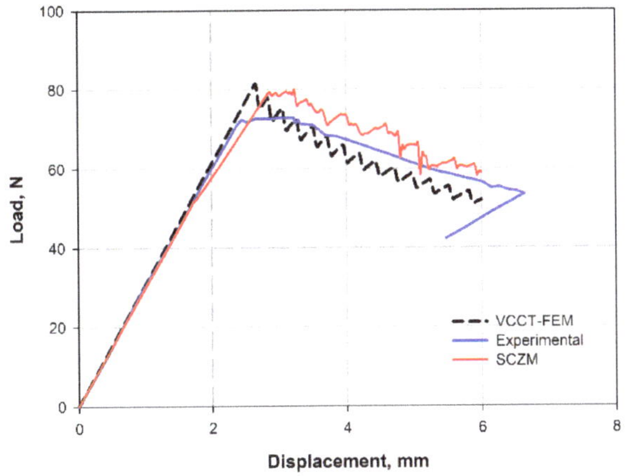

Figure 5. Load–displacement relation for DCB test.

Figure 6 shows predicted compliance using VCCT data. The compliance was calculated at each step of the curve in Figure 5 (black colored line), a crack length correction factor Δ, which can be defined as the intersect of the linear regression with the x-axis, was determined as (7.5 mm), while the experimental value was (6.57 mm) [41]. The difference was because the VCCT ignores a lot of factors in the experimental procedure such as friction between testing machine elements, interaction between layers, and some bridging in some areas during the test.

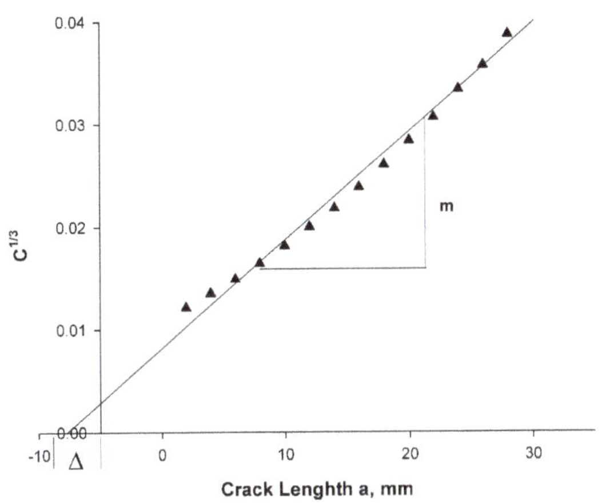

Figure 6. Compliance calibration fits to static data for DCB with Teflon insert.

Table 2 lists the comparison of the present model of Equation (13) with experimental data. It was observed that the percentage error was little, indicating that it was 0.77% and 1.5%, which means that the model described by Equation (13) is highly accurate, as demonstrated by the very small percentage errors when compared to experimental data. This suggests that the model can be used with confidence for predictions related to whatever it describes. Table 3 shows a comparison between the crack opening displacement (COD) values determined using the (VCCT) and the experimentally measured values [11]. The results show that the VCCT model provides a good approximation to the experimental values, with percentage errors ranging from 2% to 14%. The lowest error (2%) occurs at a COD of 3.56 mm, while the highest error (14%) is observed at 6 mm. The increasing trend in errors indicates that the VCCT model is reaching its limits by accurately predicting larger displacements, possibly due to nonlinear effects or variations in material behavior. These results are consistent with previous studies that emphasize the importance of validation in fracture mechanics simulations [42]. Figure 7a,b shows the presented VCCT model validation with the experimental results of two different sources, Czabaj and Ratcliffe [34] and Murri [41], respectively. Figure 7a illustrates the IM7/8552 carbon fiber-reinforced epoxy using the data obtain from VCCT-FEM with the help of Equation (2), considering the value of a crack length correction factor $\Delta = 7.5$ when the pre-crack length was by Teflon insert. The pre-crack was created by fatigue data in Figure 7b, and the VCCT COD value δ_C in Equation (2) for each load value at each slope and substituting into Equation (12). These were the two techniques, because the first technique which used the crack length correction factor misses estimating the delamination behaviors in the case of fatigue pre-crack. This can be attributed to the fatigue pre-crack being a naturally propagated crack under cyclic loading, which would give a rough, uneven crack front with some fiber bridging effects. This action introduced additional energy dissipation, whereas, for the Teflon insert pre-crack, it was perfectly sharp and clean, which would introduce a well-defined and sharp crack front. Additionally, for the case of fatigue, the softening action, residual stresses, and fiber/matrix interactions change the crack propagation characteristics [43,44]. Figure 8 shows the relation between crack length and corresponding failure mode after peak load reach, calculated using FEM associated VCCT and the analytical model presented using Equation (14). It was clearly observed that the two models give reasonably close values. The two models were validated against each other, showing good agreement. This gives confidence in the results and allows engineers to use either model for fracture toughness calculations, which are crucial for ensuring the safety and integrity of structures. The data predicted by the model obtained from Equation (13) provide reasonable results compared to the experimental work in Refs. [11,34,45]. The advantage of the present model is its simplicity and fast execution. Additionally, it can be considered a non-destructive test, requiring only the strength of the laminate in the transverse direction and the Young's modulus in the same direction. This model can be particularly useful in material selection.

Table 2. Comparison of the present model of Equation (13) with experimental data.

Laminate System	Surface Release Energy G_{IC}, kJ/m^2	Model Equation (13)	Error, %
T300/913 [45]	0.258	0.26	0.77
IM7/8552 [11]	0.27	0.26	3.7
Model [11]	0.23	0.26	11
Ref. [34]	0.24	0.27	12.5

Table 3. VCCT (COD) and peak load predicted with analytical model and experimental work.

Experimental, δ COD, mm [11]	VCCT, δ COD, mm	Error, %
2.45	2.34	4
3.56	3.5	2
3.7	3.53	5
4.2	3.8	10
6	5.14	14

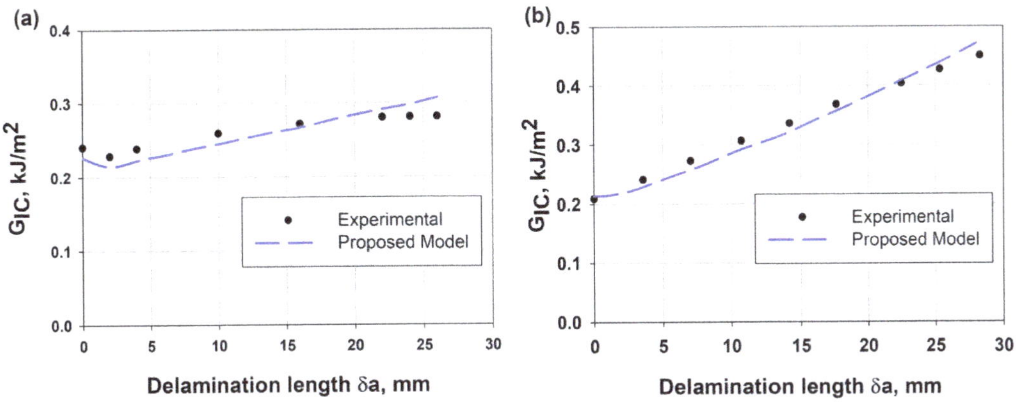

Figure 7. Resistance curve (R-curve) for IM7/8552 carbon fiber (**a**) with Teflon pre-crack and (**b**) with fatigue pre-crack compared with the experimental work of Ref. [34].

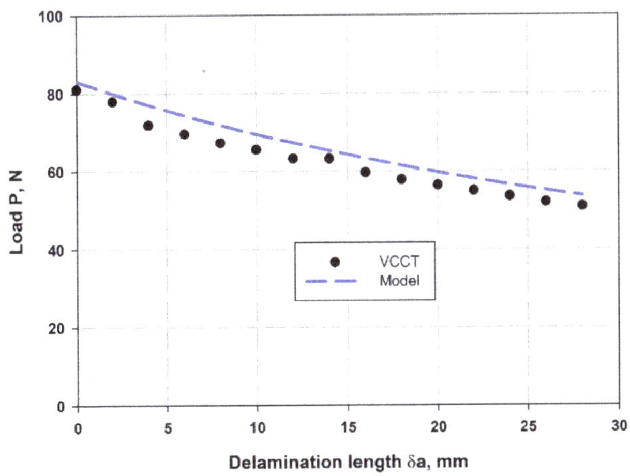

Figure 8. Load–delamination extension length relation.

6. Conclusions

Delamination is one of the most critical failure modes in composite laminates. This study can be concluded into the following items:

1. Delamination is one of the most critical failure modes in composite laminates. Accurate prediction methods are essential.
2. The finite element method (FEM) has proven to be a valuable approach in predicting delamination. In this study, FEM was integrated with the VCCT.
3. The proposed model showed good agreement with experimental results:
    a. IM7/8552 carbon fiber laminates: 0.77% error.
    b. T300/913: 1.5% error.
4. The analytical model, developed using VCCT data, demonstrated high accuracy in predicting the resistance curve and crack opening displacement (COD).
5. Errors ranged from 2% to 14%, which is acceptable from a scientific perspective.
6. The peak load after crack initiation was estimated analytically with high accuracy.

Author Contributions: Conceptualization, M.Y.A. and A.F.M.; methodology, M.Y.A. and M.K.H.; software, M.Y.A. and A.H.B.; validation, M.Y.A.; formal analysis, M.Y.A. and A.F.M.; investigation, M.Y.A.; resources, A.H.B.; writing—original draft preparation, M.Y.A.; writing—review and editing, M.K.H.; supervision, M.Y.A.; project administration, M.Y.A.; funding acquisition, A.F.M. All authors have read and agreed to the published version of the manuscript.

Funding: This research work was funded by Umm Al-Qura University, Saudi Arabia, under grant number 25UQU4290444GSSR01.

Data Availability Statement: The original contributions presented in this study are included in the article. Further inquiries can be directed to the corresponding author.

Acknowledgments: The authors extend their appreciation to Umm Al-Qura University, Saudi Arabia, for funding this research work through grant number 25UQU4290444GSSR01.

Conflicts of Interest: The authors declare no conflicts of interest.

References

1. Hassan, M.K.; Mohammed, Y.; Salem, T.; Hashem, A. Prediction of nominal strength of composite structure open hole specimen through cohesive laws. *Int. J. Mech. Mech. Eng. IJMME-IJENS* **2012**, *12*, 1–9.
2. Chau-Dinh, T.; Zi, G.; Lee, P.-S.; Rabczuk, T.; Song, J.-H. Phantom-node method for shell models with arbitrary cracks. *Comput. Struct.* **2012**, *92–93*, 242–256. [CrossRef]
3. Nguyen-Thanh, N.; Valizadeh, N.; Nguyen, M.N.; Nguyen-Xuan, H.; Zhuang, X.; Areias, P.; Zi, G.; Bazilevs, Y.; De Lorenzis, L.; Rabczuk, T. An extended isogeometric thin shell analysis based on Kirchhoff–Love theory. *Comput. Methods Appl. Mech. Eng.* **2015**, *284*, 265–291. [CrossRef]
4. Nilsson, K.-F.; Giannakopoulos, A.E. A finite element analysis of configurational stability and finite growth of buckling driven delamination. *J. Mech. Phys. Solids* **1995**, *43*, 1983–2021. [CrossRef]
5. Robinson, P.; Javidrad, F.; Hitchings, D. Finite element modelling of delamination growth in the DCB and edge delaminated DCB specimens. *Compos. Struct.* **1995**, *32*, 275–285. [CrossRef]
6. Fleming, D.C. Delamination Modeling of Composites for Improved Crash Analysis. *J. Compos. Mater.* **2001**, *35*, 1777–1792. [CrossRef]
7. La Saponara, V.; Muliana, H.; Haj-Ali, R.; Kardomateas, G.A. Experimental and numerical analysis of delamination growth in double cantilever laminated beams. *Eng. Fract. Mech.* **2002**, *69*, 687–699. [CrossRef]
8. Camanho, P.P. *Numerical Simulation of Delamination Growth in Composite Materials*; NASA Langley Technical Report Server: Hampton, VA, USA, 2001.
9. Roudolff, F.; Ousset, Y. Comparison between two approaches for the simulation of delamination growth in a D.C.B. specimen. *Aerosp. Sci. Technol.* **2002**, *6*, 123–130. [CrossRef]
10. Rybicki, E.F.; Kanninen, M.F. A finite element calculation of stress intensity factors by a modified crack closure integral. *Eng. Fract. Mech.* **1977**, *9*, 931–938. [CrossRef]
11. Abdellah, M.Y. Delamination modeling of double cantilever beam of unidirectional composite laminates. *J. Fail. Anal. Prev.* **2017**, *17*, 1011–1018. [CrossRef]
12. Elder, D.J.; Thomson, R.S.; Nguyen, M.Q.; Scott, M.L. Review of delamination predictive methods for low speed impact of composite laminates. *Compos. Struct.* **2004**, *66*, 677–683. [CrossRef]
13. Geng, D.; Liu, Y.; Shao, Z.; Lu, Z.; Cai, J.; Li, X.; Jiang, X.; Zhang, D. Delamination formation, evaluation and suppression during drilling of composite laminates: A review. *Compos. Struct.* **2019**, *216*, 168–186. [CrossRef]
14. Lu, X.; Ridha, M.; Chen, B.Y.; Tan, V.B.C.; Tay, T.E. On cohesive element parameters and delamination modelling. *Eng. Fract. Mech.* **2019**, *206*, 278–296. [CrossRef]
15. Kumar, D.; Roy, R.; Kweon, J.-H.; Choi, J.-H. Numerical Modeling of Combined Matrix Cracking and Delamination in Composite Laminates Using Cohesive Elements. *Appl. Compos. Mater.* **2016**, *23*, 397–419. [CrossRef]
16. Marjanović, M.; Meschke, G.; Vuksanović, D. A finite element model for propagating delamination in laminated composite plates based on the Virtual Crack Closure method. *Compos. Struct.* **2016**, *150*, 8–19. [CrossRef]
17. Krueger, R. 1—The virtual crack closure technique for modeling interlaminar failure and delamination in advanced composite materials. In *Numerical Modelling of Failure in Advanced Composite Materials*; Camanho, P.P., Hallett, S.R., Eds.; Woodhead Publishing: Wimblington, UK, 2015; pp. 3–53. [CrossRef]
18. Li, D.H. Delamination and transverse crack growth prediction for laminated composite plates and shells. *Comput. Struct.* **2016**, *177*, 39–55. [CrossRef]

19. Whitcomb, J.D. Three-Dimensional Analysis of a Postbuckled Embedded Delamination. *J. Compos. Mater.* **1989**, *23*, 862–889. [CrossRef]
20. Chen, H.-P. Shear deformation theory for compressive delamination buckling and growth. *AIAA J.* **1991**, *29*, 813–819. [CrossRef]
21. Hitchings, D.; Robinson, P.; Javidrad, F. A finite element model for delamination propagation in composites. *Comput. Struct.* **1996**, *60*, 1093–1104. [CrossRef]
22. Aymerich, F.; Lecca, G.; Priolo, P. Modelling of delamination growth in composite laminates by the virtual internal bond method. *Compos. Part A Appl. Sci. Manuf.* **2008**, *39*, 145–153. [CrossRef]
23. Mohammed, Y.; Hassan, M.K.; Hashem, A. Analytical model to predict multiaxial laminate fracture toughness from 0 ply fracture toughness. *Polym. Eng. Sci.* **2014**, *54*, 234–238. [CrossRef]
24. Abdellah, M.Y.; Hassan, M.K.; Mohamed, A.F.; Khalil, K.A. A novel and highly effective natural vibration modal analysis to predict nominal strength of open hole glass fiber reinforced polymer composites structure. *Polymers* **2021**, *13*, 1251. [CrossRef]
25. van der Meer, F.P.; Moës, N.; Sluys, L.J. A level set model for delamination—Modeling crack growth without cohesive zone or stress singularity. *Eng. Fract. Mech.* **2012**, *79*, 191–212. [CrossRef]
26. Krueger, R. Virtual crack closure technique: History, approach, and applications. *Appl. Mech. Rev.* **2004**, *57*, 109–143. [CrossRef]
27. Jokinen, J.; Kanerva, M.; Wallin, M.; Saarela, O. The simulation of a double cantilever beam test using the virtual crack closure technique with the cohesive zone modelling. *Int. J. Adhes. Adhes.* **2019**, *88*, 50–58. [CrossRef]
28. Abdellah, M.Y.; Gelany, A.F.; Mohamed, A.F.; Khoshaim, A.B. Protection of limestone Coated with Different Polymeric Materials. *Am. J. Mech. Eng.* **2017**, *5*, 51. [CrossRef]
29. Abdellah, M.Y. Ductile Fracture and S–N Curve Simulation of a 7075-T6 Aluminum Alloy under Static and Constant Low-Cycle Fatigue. *J. Fail. Anal. Prev.* **2021**, *21*, 1476–1488. [CrossRef]
30. Abdellah, M.Y. Essential Work of Fracture Assessment for Thin Aluminium Strips Using Finite Element Analysis. *Eng. Fract. Mech.* **2017**, *179*, 190–202. [CrossRef]
31. Wanthal, S.; Schaefer, J.; Justusson, B.; Hyder, I.; Engelstad, S.; Rose, C. Verification and validation process for progressive damage and failure analysis methods in the NASA Advanced Composites Consortium. In Proceedings of the American Society for Composites (ASC) Technical Conference, West Lafayette, IN, USA, 22–25 October 2017.
32. Alharthi, H.; Abdellah, M.Y. Stress Analysis and Strength Prediction of Carbon Fiber Composite Laminates with Multiple Holes Using Cohesive Zone Models. *Polymers* **2025**, *17*, 124. [CrossRef]
33. Alssayegh, A.; Abdellah, M.Y.; Hassan, M.K.; Azam, S.; Melaibari, A.; Khashaba, U. Optimizing high cycle fatigue predictions in notched Al 7075-T6: An analytical approach to rotating bending behavior. *Results Eng.* **2025**, *25*, 103623. [CrossRef]
34. Czabaj, M.W.; Ratcliffe, J.G. Comparison of intralaminar and interlaminar mode I fracture toughnesses of a unidirectional IM7/8552 carbon/epoxy composite. *Compos. Sci. Technol.* **2013**, *89*, 15–23. [CrossRef]
35. Librescu, L.; Song, O. *Thin-Walled Composite Beams: Theory and Application*; Springer Science & Business Media: Berlin/Heidelberg, Germany, 2005; Volume 131.
36. Wang, C.H. *Introduction to Fracture Mechanics*; DSTO Aeronautical and Maritime Research Laboratory: Melbourne, Australia, 1996.
37. Song, W.; Chen, Y.; Mu, Z.; Wang, Y.; Zhang, Z.; Wang, Z.; Liu, L.; Zhang, B.; Li, Y.; Li, B. A feather-inspired interleaf for enhanced interlaminar fracture toughness of carbon fiber reinforced polymer composites. *Compos. Part B Eng.* **2022**, *236*, 109827. [CrossRef]
38. Dávila, C.G.; Rose, C.A.; Camanho, P.P. A procedure for superposing linear cohesive laws to represent multiple damage mechanisms in the fracture of composites. *Int. J. Fract.* **2009**, *158*, 211–223. [CrossRef]
39. Budiansky, B.; Hutchinson, J.W.; Evans, A.G. Matrix fracture in fiber-reinforced ceramics. *J. Mech. Phys. Solids* **1986**, *34*, 167–189. [CrossRef]
40. Reeder, J.R.; Crews, J.H., Jr. Mixed-mode bending method for delamination testing. *AIAA J.* **1990**, *28*, 1270–1276. [CrossRef]
41. Murri, G.B. Evaluation of delamination growth characterization methods under mode I fatigue loading. In Proceedings of the 15th US-Japan Conference on Composite Materials, Arlington, TX, USA, 1–3 October 2012.
42. Anderson, T.L.; Anderson, T.L. *Fracture Mechanics: Fundamentals and Applications*; CRC Press: Boca Raton, FL, USA, 2005.
43. Khan, R. Fiber bridging in composite laminates: A literature review. *Compos. Struct.* **2019**, *229*, 111418. [CrossRef]
44. Krueger, R. *An Approach to Assess Delamination Propagation Simulation Capabilities in Commercial Finite Element Codes*; NASA Langley Technical Report Server: Hampton, VA, USA, 2008.
45. Pinho, S.T. *Modelling Failure of Laminated Composites Using Physically-Based Failure Models*; University of London: London, UK, 2005.

polymers

MDPI

Article

Influence of CNT Filler in Polymer Matrix on Impact Damage Propagation in the Volume of Carbon Fiber Laminates

Egor Morokov [1,*], Pavel Shershak [2], Mikhail Burkov [3], Alexander Eremin [3], Elizaveta Popkova [2], Nikolay Yakovlev [2] and Irina Zhiltsova [1]

1 Emanuel Institute of Biochemical Physics, Russian Academy of Sciences, 119334 Moscow, Russia; iraida_n@mail.ru

2 Federal State Unitary Enterprise, All-Russian Scientific Research Institute of Aviation Materials, National Research Center "Kurchatov Institute", 105005 Moscow, Russia; tweek@list.ru (P.S.); lab30@viam.ru (E.P.); nick_62@mail.ru (N.Y.)

3 Institute of Strength Physics and Materials Science of Siberian Branch of Russian Academy of Sciences, 634055 Tomsk, Russia

* Correspondence: es_morokov@yahoo.com

Abstract: The addition of nano-sized fillers into the polymer matrix of carbon fiber laminates is considered today as one of the ways of increasing the strength and resistance of the material to mechanical loads. The paper considers the effect of the addition of single-walled carbon nanotubes (CNTs) on the development of impact damage in laminates. Studies of the volume microstructure and its damage were carried out using high-resolution ultrasound imaging. The effect of damage propagation in laminates with an increase in the concentration of CNTs from 0 to 0.5 wt% was shown. The addition of CNTs decreased the area of damage in the upper and lower part of laminates but increased the area of damage in the middle plies. The results were discussed in combination with data on impact histories of laminates.

Keywords: polymer composite; CFRP; matrix modification; impact damage; ultrasound; volume imaging; CNT

1. Introduction

Carbon fiber reinforced polymer (CFRP) composites are a promising class of materials with outstanding properties such as strength, stiffness, and lightness. These materials find a wide application in various fields, including aerospace, sports equipment, automotive, and others [1,2]. However, CFRP laminate is prone to delamination under mechanical loads due to the low interlaminar fracture toughness. This is primarily determined by properties of the matrix and the fiber–polymer bond. One of the current challenges in the field of composites is to improve interaction between the polymer matrix and reinforcing fibers, increase the fracture toughness, and ultimately improve the crack resistance.

In recent decades, promising studies have been conducted to improve fracture toughness by modifying the polymer matrix [3–5], modifying the surface of reinforcing fibers [6–8], or by vertically crosslinking (pinned) carbon layers [9,10]. Matrix modification was carried out by incorporating additional nanoparticles into the polymer matrix, namely: carbon multi-walled carbon nanotubes (MWCNT) [3,11–14], graphene oxide (GO) [15], graphene nanoplatelets (GNP) [16,17], nanoscale clay (NC) [18], short aramid fibers [19,20], and various combinations of them [21,22]. The works showed that the addition of these nanoparticles led to improvements in interlaminar strength. Nanophase evokes the concept of multiscale reinforcement, which connects the boundary layers vertically, leading to the redistribution and dissipation of energy during destruction. Ou et al. [13] demonstrated that the crack front propagated alternately above and below the nanotube-reinforced interlayer. This significantly improves the fracture toughness of the laminates. The mechanism is highly dependent on how the nanotubes were integrated into the matrix. They suggested [13] that directly depositing CNTs on plies as low-density thin layers resulted in significant improvements in interlaminar properties, while mixing the MWCNT with the matrix leads to agglomeration and defects. Liu et al. [21] incorporated graphene oxide and MWCNTs into composite laminates, achieving significant improvement in interlaminar fracture toughness. Koirala et al. [23] used ultra-thin aligned CNT sheets (100 nm) to improve the interlaminarity without changing the weight or thickness of laminates. These studies confirm the positive effect of CNTs on interlaminary fracture toughness. Tehrani et al. [24] showed that the incorporation of MWCNTs into CFRP increased the impact energy absorption by up to 21%. Bedsole et al. [11] showed that the introduction of nanotubes resulted in improvements in fracture toughness under both quasi-static (+34%) and dynamic (+16%) loading conditions. Alshehri et al. [3] found that the lowest delamination initiation force was observed in impact tests for laminates with a pure polymer matrix, and it increased with increasing concentrations of MWCNTs at 0.2 and 0.4 wt% in the polymer. Despite the direct relationship between the concentration of nanofillers in a matrix and their fracture toughness, there is a limit on nanophase concentration. A high affinity of MWCNT can lead to the agglomeration of nanoparticles, reducing the effect of introducing them. In addition, when a filler is introduced, the viscosity of resin increases, affecting the diffusion and impregnation of prepreg during manufacturing. To reduce particle agglomerate, chemical functionalization of nanofillers was used, significantly affecting the dispersion of nanofiller in the matrix [25]. Another possible solution was layer-by-layer application using a spray [21].

Interlayer fracture toughness is especially important for low-velocity impacts, when the main energy of mechanical action is concentrated in the interlayer areas. Under such loads, so-called barely visible damages are formed in the CFRP volume. These damages are not visible from the surface, but they can occupy a large area in the composite volume and significantly reduce its load-bearing capacity. The study of the microstructure of CFRP under low-velocity impact has been challenging since most damage appears in the volume of optically opaque laminate. Therefore, the application of non-destructive techniques for characterizing CFRP looks attractive. Today, quantitative characterization and visualization of impact damage in CFRP can be performed using X-ray [26–29] and ultrasound [3,30–33] techniques or thermography [34,35]. The non-destructive approaches solve two problems: the validation of numerical models for predicting impact damage [36–42], and the monitoring of irreversible changes to the microstructure without cutting and polishing composite samples. Direct experimental results from the non-destructive observation of structural disturbances significantly improve current and future numerical models of CFRP failure prediction. Therefore, in this work, the high-resolution ultrasound visualization technique is used to observe the damage in CFRP reinforced with SWCNTs under low-velocity impact. The features of damage appearance and location in composites with different CNT concentrations are revealed and discussed.

2. Materials and Methods

2.1. Specimen Preparation

Composites were prepared using biaxial fabric CBX300 (Mitsubishi Pyrofil TR50S 12K, Sacramento, CA, USA) and epoxy binder R&G Epoxy L (GL2 hardener) with the addition of single-walled carbon nanotubes (SWCNT) TUBALL (OcSiAl, Luxembourg), supplied by the manufacturer in the form of a pre-dispersed composition of CNT with ethoxylated alcohol. SWCNTs had a mean diameter of 1.6 ± 0.4 nm and a length of >5 μm.

This CNT batch was added to the epoxy resin and thoroughly mixed before the hardener was introduced. The batch was mechanically mixed with epoxy for 10–20 min and a simple visual homogeneity test was applied, both proposed by the manufacturer. Carbon fabrics were laid out using manual layup (stacking sequence $[+45/-45/0/90]_{5S}$) followed by vacuum bagging and pressing in a thermal press Gotech 7014 under a pressure of 0.7 MPa and a temperature of 80 °C. After a dwell time of one hour, the laminate was removed from the mold, the release film and breather fabric were removed, and post-curing for 24 h at a temperature of 80 °C was performed in a heating chamber. After the post-curing, the specimens for testing were cut on a CNC milling machine Purelogic RM0813 (Voronezh, Russia) with polycrystalline diamond mill with water cooling. One unmodified epoxy laminate and four laminates modified with 0.1, 0.2, 0.3, and 0.5 wt% of SWCNT were manufactured for testing. The specimen geometry of 100 × 150 mm was based on the ASTM D7136 standard [43]. The thickness of laminates was 4.8 mm.

2.2. Scanning Electron Microscopy

The microstructure was evaluated using a Quanta 200 3D scanning electron microscope (FEI Company, Hillsboro, OR, USA) (SEM). Preliminary samples of 4 × 4 × 4.8 mm were cut from the laminates and heated in a NETZSCH STA409PC/PG (Selb, Germany) thermogravimetric analysis (TGA) unit at the following conditions: heating from ambient temperature to 873 K at a rate of 15 K/min with an Argon flow rate of 50 mL/min. TGA procedure leads to the decomposition and evaporation of the epoxy matrix while leaving the carbon fibers and CNTs undamaged. Subsequent SEM analysis of the TGA residue allows to visualize CNT distribution.

2.3. Mechanical Tests

Low-velocity impact tests were carried out using the Drop Weight Tester HIT230F from Zwick Roell (Zwick GmbH Co.KG, Ennepetal, Germany). The diameter of the impactor was 25.4 mm (1 in). The impact energy was ~30 J with a mass of 5.6 kg.

2.4. High-Resolution Ultrasound Imaging

The scanning impulse acoustic microscope SIAM-2011 [30,44] developed and produced by the Institute of Biochemical Physics, Russian Academy of Science (Moscow, Russia), has been applied in experiment. An acoustic lens with a frequency of 50 MHz and angle aperture of 11° provides a lateral resolution of 60 μm in the CFRP volume at the depth of a few mm. Short probe pulses of 25 ns in duration provide an axial resolution of 40 μm. Ultrasound devices can visualize extended cracks and delaminations with a nanoscale gap. The reflection of these defects is complete, making it possible to detect and visualize the damage in a laminated volume. The volume microstructure of laminates can be represented as a set of B-scans and C-scans taken at different depths inside the specimen, which depict the microstructure in vertical and horizontal sections of the laminate, respectively. A detailed description of visualization modes, principles of formation, and interpretation of carbon fiber elements are presented in the work [45].

3. Results

The results of the research are the comparison of the impact history values (loads, energies) and data of high-resolution ultrasound visualization of damage in the volume of laminates. A direct comparison of the values of impact loads and energies with the sizes of damage and their localization over the laminate thickness will allow a better understanding of the mechanism of the influence of CNT on damage propagation.

3.1. SEM Investigations

Figure 1 shows SEM micrographs of CNT distribution in laminates. The CNTs were located between carbon fibers and formed cross-links between them. The distribution of CNTs was irregular, with particles congregating and forming a continuous transversal network of group CNTs throughout the laminates. The network increased with increasing filler concentration. Large conglomerates of CNTs surrounded several carbon fibers at concentrations of 0.3 wt% and 0.5 wt%.

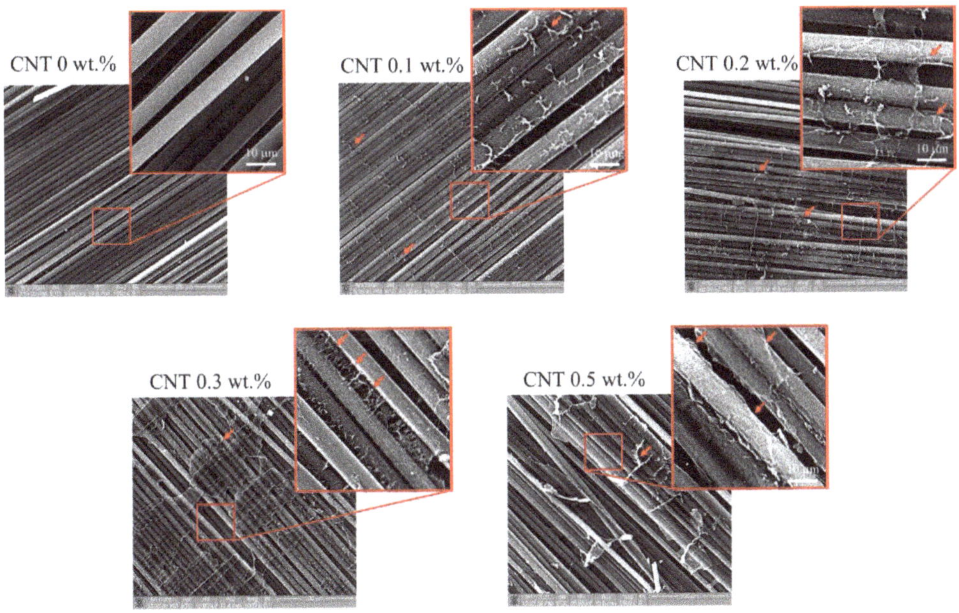

Figure 1. SEM micrographs of laminates with various concentrations of CNT. The CNT conglomerates formed a continuous network linking carbon fibers in the transverse direction. The size of CNT meshes increased with increasing of concentration. Arrows are the CNT conglomerates.

3.2. Impact Behavior

The force-time histories obtained from the impacts on the laminates are shown in Figure 2. The force-time dependencies of all the impact tests were analyzed in order to evaluate the delamination threshold load F_d, and peak force F_{max} (Table 1). All the specimens had approximately equal values for the threshold load F_d of ~4 kN and the peak force F_{max} of ~11.5 kN. The lowest value of F_{max} of 10.5 kN occurred in laminates containing 0.3 wt% of CNT. It is clear that the higher initial energy E_{ini} at force of damage initiation F_{d} will result in the lower energy for damage propagation and the lower damages in the laminates. The laminate with 0.3 wt% of CNTs had the higher value of the energy E_{ini}; however, the peak force F_{max} was slightly lower than other specimens. Despite the similar force histories (Figure 2a), the curves of the laminates with 0.2 wt% and 0.3 wt% of CNTs showed a significant drop of load immediately after the peak forces. At the same time, the total displacement of the impactor was about 4.6 mm (Figure 2c). It can be seen that there was an elastic energy component in the energy-time plot (Figure 2b) and that the absorbed energy E_{ab} increased with increasing of CNT concentration (Table 1); the rebound of the impactor also occurred, as indicated by the curves of force versus impactor displacement (Figure 2c).

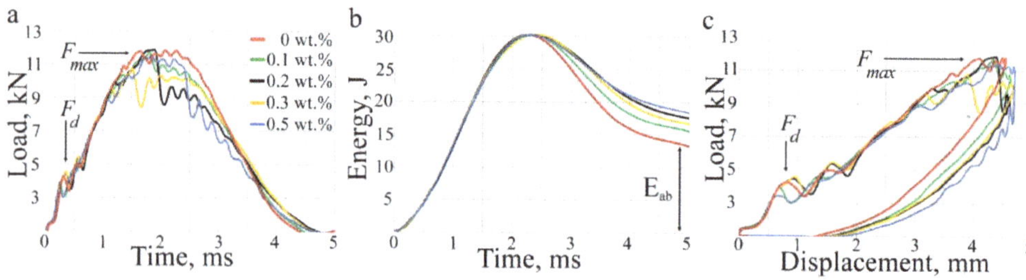

Figure 2. Mechanical test results. Impact force histories (**a**), impactor energy history (**b**), and curves of force versus impactor displacement (**c**). The threshold load of delamination F_d has minimal variability and is equal to 4 kH (**a**). The peak load F_d occurred at the indenter displacement of 0.6 ÷ 0.9 mm (**c**), which is associated with the appearance of the first damages. F_{max} is the peak force that occurred during the impact; the values of F_{max} were about 11 kH. The total impact energy was 30 J; however, only part of it was absorbed by the laminates. The absorbed energy E_{ab} (**b**) increased with increasing CNT concentration.

Table 1. Impact data on delamination threshold load and peak force of the specimens.

Specimen	0 wt%	0.1 wt%	0.2 wt%	0.3 wt%	0.5 wt%
Threshold load F_d, kN	4.25	3.90	4.44	4.55	3.40
Peak force F_{max}, kN	11.80	11.68	11.84	10.80	11.42
Initiation Energy E_{ini}, J	2.10	1.70	2.70	2.80	1.80
Energy to peak force E_{max}, J	29.73	27.62	27.98	23.22	27.85
Absorbed Energy E_{ab}, J	13.50	15.50	17.30	16.40	18.20

3.3. Projected Damaged Areas of the Laminates

Impact damage zones in the volume were assessed by visualizing the microstructure over the thickness of laminates. Figure 3 displays projected damaged areas of the laminates at different depths below the impacted surface. Damages are displayed as bright areas on a dark background. It was found that the damage occurred unevenly across the thickness. Interplay delaminations were found in three zones (Figure 3): upper, middle and lower, which corresponded to a depth range of 0.4 ÷ 0.8 mm, 2.0 ÷ 2.8 mm and 3.3 ÷ 4.4 mm (left, central, and right column, respectively) with some variations. For each type of laminate, damages in the upper zone close to the impacted surface are characterized by a minimal size of delaminations that appear from the third up to the seventh interplay (left column). Damages in the middle of the laminates have a maximal square and most of them are located around the central 90–90° interface between 17th and 22nd plies. Delaminations in the lower part of the laminates located from the 27th to 37th interfaces decrease in size as depth increases (right column).

Figure 4 shows the histograms of the correlation between the damaged areas in three regions and CNT concentration. Based on ultrasound data, it can be concluded that the addition of CNTs to the matrix has two distinct effects. First, it reduces the damaged areas in the upper and lower parts of the laminates. Second, it increases the central damaged area. The central damage of the laminates with CNTs extends to the edges, which can affect the spread of delaminations, while the laminate without filler seals delaminations that can affect the size of the damage located in the lower part of the laminate. The total area of delamination observed in the volume of laminates increases with increasing concentration of CNTs; however, this does not mean a decrease in the residual strength of the laminates, which depends on the direction of damaged layers and broken fibers, as well as on the location of delamination in the thickness of the laminate.

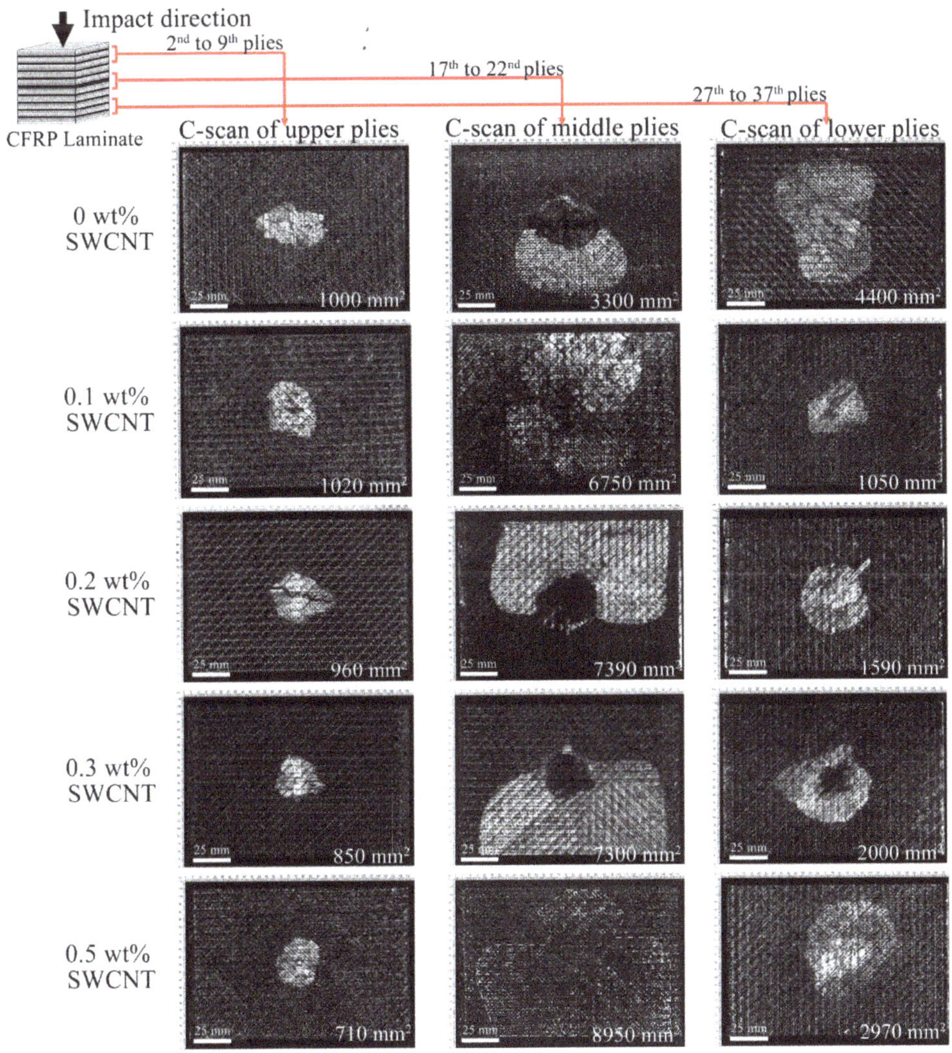

Figure 3. C-scans of the impact damage in the volume of the laminates with various concentrations of CNTs. The images show three zones in the thickness of the laminates with the calculated area of damage. Bright areas in the C-scans correspond to interlayer delaminations. The size of damages in the upper plies (left column) decreased with increasing of CNT concentration. Projected damaged areas in the middle zone (central column) of the laminates increased with increasing of CNT concentration. The laminate with a CNT content of 0.1 wt% has minimal damage in the lower plies (right column); the damage area increases with increasing concentration. The maximum size of delaminations (4400 mm^2) was observed in the pure laminate.

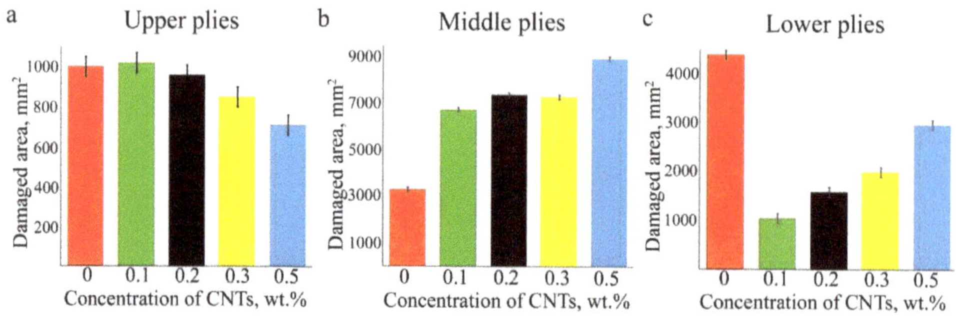

Figure 4. Diagrams showing the distribution of damage zones in laminates depending on CNT concentration. The addition of CNTs reduces the damage in the top (**a**) and bottom (**c**) plies, while increasing it in the middle (**b**) plies. The spread of values relates to the method used for calculating the damaged area in C-scans.

3.4. *Layer-by-Layer Imaging of Damage of the Laminates*

Layer-by-layer ultrasonic imaging of the damage depicts the delamination that appeared simultaneously in each interface of the damage zones. Figure 5 shows the sequences of damage propagation in the upper plies of the laminates. The interlayer delaminations are displayed as bright areas against the dark background that depicted fiber orientations in adjacent plies. Damage occurs at the 2–3 interface and below; delaminations spread along the fiber orientation of neighboring plies of the interfaces. Usually, the laminates have double triangle-shaped delaminations symmetric to the impact point. The exception is the 0–90° boundary, where the front of the delamination development has a rectangular shape. The development of delaminations from the impacted surface comes about due to the rotation of damaged areas, with jumps from one interface to another. It was also found that the laminates with 0.1 and 0.5 wt% of CNTs had a conglomeration of fillers that can be seen in the C-scans as bright small-scaled elements (marked with a star in Figure 5). Therefore, the ultrasound contrast decreased close to the middle zone of the laminates (Figure 3). Nevertheless, the inhomogeneous distribution of CNTs did not affect the impact resistance. The size of the delaminations in the upper plies decreased with increasing CNT concentration.

On the other hand, the damage in the middle plies had the largest area (Figure 4), and the total area increased with the increase of CNT concentration. These damages were also distributed across multiple interfaces. Figure 6 shows the damage distribution in the middle of the laminate with 0.3 wt% of CNT. A similar damage propagation mechanism was observed in the other laminates. The mechanism was that the damage appeared step by step (B-scan, Figure 6) at the four interfaces, and the largest delamination occurred at the 20–21 interface (90–90° layers). The smallest delamination was observed between the 0° and 90° layers (19–20 interface), which can be explained by the largest angular differences in fiber orientations. The distribution of damage across the interfaces ensured the integrity of the laminates and their residual strength.

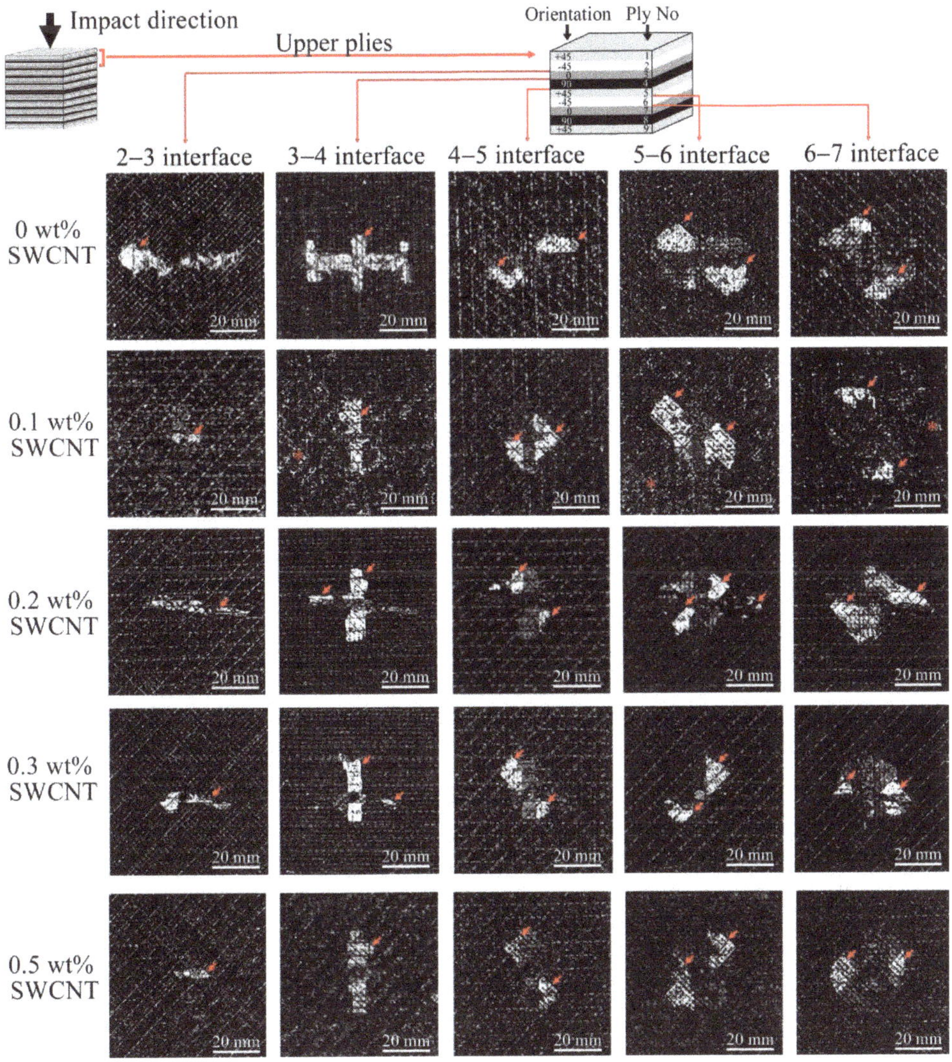

Figure 5. Layer-by-layer C-scans of damages in upper plies of impacted laminates with various CNT concentrations (from up to down). The size of damage decreases with increasing concentration of CNTs. Arrows depict the interlayer delaminations, stars are CNT conglomerates clearly observed in the laminate with 0.1 wt% of CNTs. The shape of the delaminations is determined by the orientation of the fibers in adjacent plies. Delaminations appeared in each interlayer from the second to the seventh.

The damages in the lower plies of the laminates appeared from the 33rd ply up to the back surface of the laminates. Figure 7 shows the damage occurrence in the laminate with 0.2 wt% of CNTs. The delaminations have the shape of a symmetrical triangle with the maximum area at the interfaces from the 33rd to the 36th layers; then the size of the damage decreases, with delaminations observed up to the 39th ply. A similar damage propagation was observed in the other laminates with the maximal delaminations close to the 30th ply. The shape and front of the delaminations are determined by the orientations of the fiber in neighboring plies.

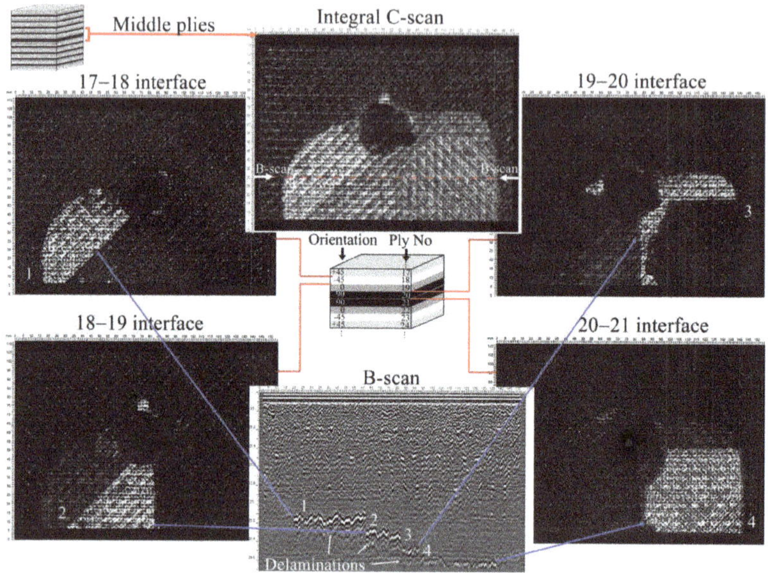

Figure 6. Layer-by-layer C-scans of damages in the middle plies of the laminate with 0.3 wt% of CNTs. Damages appeared step by step in four interfaces that were seen in the B-scan. The position of the B-scan is marked in the integral C-scan showing the total amount of damage in the middle zone. The shape of the delaminations is determined by the orientation of adjacent plies. The orientation and numbering of the plies are shown in the scheme. C-scans show a damage distribution typical for all tested laminates.

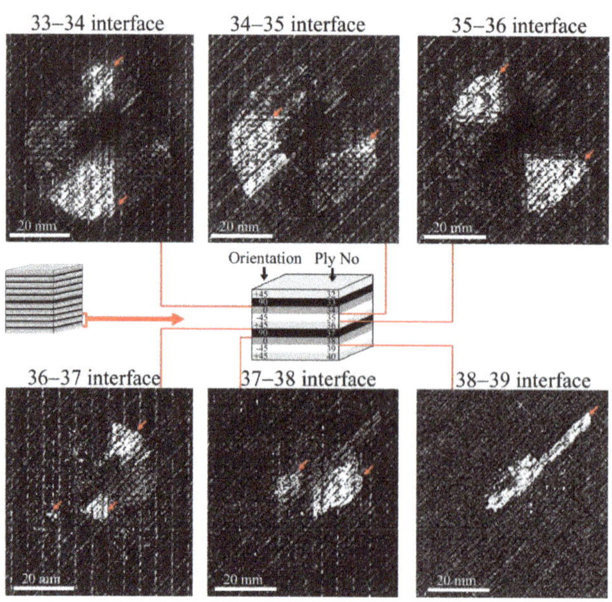

Figure 7. Layer-by-layer C-scans of damages in the lower plies of the laminate with 0.2 wt% of CNTs. Maximal delaminations formed close to the 30 ply and decreased towards the back surface of the laminate. The shape of the delaminations is determined by the orientation of adjacent plies. The orientation and numbering of the plies are shown in the scheme. C-scans show a damage distribution typical for all tested laminates with CNT filler. Arrows are the interlayer delaminations.

4. Discussion

The main action of nanofillers in the CFRP laminates is to inhibit crack growth between the fibers and the matrix, which leads to effective energy absorption. It is known that improvements in mechanical properties and efficient load transfer from the fibers to the matrix depend on the interfacial adhesion between CFRP components [12,46–48]. The inclusion of nanoparticles causes a change in interlayer interaction to intralayer, which results in an increase in the resistance of the matrix to cracking. In this work, it was found that the efficiency in facilitating the impact energy transfer between the fiber and matrix increases with increasing CNT concentration. The area of damage in the upper part of the laminates decreased (Figure 3). CNTs embedded between carbon fibers (Figure 1) can increase the interactions between plies along the direction of impact (vertically) and delay interlayer delaminations in the upper zone of laminates. The increased interaction between plies in laminates with CNT results in the transfer of impact energy from the upper plies to the middle region, where maximum stress was caused by the accumulation of energy; its abrupt release led to extended interlayer delaminations.

Based on the obtained ultrasonic data, the scheme of delamination location can be summarized as follows. The damage appeared from the second layer under the impactor. The size of the delamination increased from interface to interface up to the 8th ply of the laminates, after which the damage was not visible until the 17th ply, where the maximum delaminations formed step by step on the four lower interfaces. Further damage was observed only under the 30th ply, where delaminations spread from ply to ply in decreasing size. In combination with the impact history, it can be summarized that the damage initiated at the impactor's deflection, about 0.7 mm (Figure 2c), corresponded to the depth of damages in the upper plies seen in the C-scans. (Figure 3). Similar damage initiation energy in all laminates did not affect the occurrence or propagation of damage in the upper layers of the laminates. Moreover, slightly lower E_{ini} values (Table 1) were found in laminates with 0.1 and 0.5 wt% of CNTs. This might be due to the cluster distribution of the filler (Figure 3) which could affect the redistribution of stress in the volume of the material. In addition, a correlation between the absorbed energy E_{ab} (Figure 2b) and the integral area of damage was clearly traced, which had a dependence close to linear. An increase in the filler concentration led to an increase in the absorbed energy from 13.5 to 18.2 J. At the same time, the sum of damaged area observed in the upper, middle, and lower plies of the laminates with various CNT concentrations (Figure 4) successively increased from 8700 to 12,600 mm^2. It should be noted that the damage size increased in the middle layers of the laminates with filler (Figure 4), thus it can be assumed that most of the absorbed energy was spent on the middle plies. The addition of CNTs into the polymer matrix significantly affected the damage in the upper and lower plies of laminates, the size of which was two times smaller compared to damage in a laminate without CNTs (Figure 4). In addition, damage in laminates without CNTs had a larger spread of damaged interfaces in the region of the lower plies. Typically, these outer upper and lower areas of laminates influenced the residual strength of the material [49].

Thus, all parameters of damage in the volume of laminates, namely: the size of delaminations, their location and depth, the sequence of damage propagation, the orientation of the fibers of the damage layers, etc., can affect the future stability of laminates, which requires additional experiments to determine the correlation between the distribution of damage in the volume of laminates with CNTs and their residual compressive strength after impact.

5. Conclusions

In the work, the low-velocity impact damage appearance and propagation in the biaxial fabric laminates with various concentrations of CNTs in a polymer matrix was investigated. It was found that CNTs clustered together and formed a continuous transverse network between carbo fibers, and the size of the clusters increased with increasing CNT concentrations. Visualization of the damages in the volume microstructure was carried out using a high-frequency ultrasound imaging method, which allowed layer-by-layer localization and observation of the distribution of interlayer delaminations. Based on the data obtained, the following can be concluded:

- The addition of CNTs to the matrix reduces the damaged areas in the outer plies of the laminates; the size of damages was two times smaller compared to damage in a laminate without CNTs.
- Increasing the concentration of CNTs increases interlayer interaction and impact energy transfer through the thickness, which affects the energy absorption in laminates and increases damage in the middle region.
- CNT conglomeration affects the stress distribution in laminates, which reduces the threshold load for delamination.

We suppose that the low damages in outer upper and lower areas of laminates will influence the residual strength of the material both under static and cyclic loads.

In the future, it seems attractive to continue the work on identifying impact damage in CFRP with nano-sized filler and determining the dependence of localization and size of damages on the residual strength of laminates.

Author Contributions: Conceptualization, E.M.; methodology, E.M, M.B. and N.Y.; investigation, E.M., A.E., E.P. and I.Z.; writing—original draft preparation, E.M., P.S. and M.B.; writing—review and editing, E.M.; visualization, I.Z.; project administration, E.M.; funding acquisition, E.M. All authors have read and agreed to the published version of the manuscript.

Funding: This work was financially supported by Grant No. 24-72-10067 from the Russian Science Foundation.

Institutional Review Board Statement: Not applicable.

Data Availability Statement: Data is contained within the article.

Acknowledgments: The authors thank Sophy Melange for editing the English grammar.

Conflicts of Interest: The authors declare no conflicts of interest.

Abbreviations

The following abbreviations are used in this manuscript:

CFRP	Carbon fiber reinforced polymer
CNT	Carbon nanotube

References

1. Xu, X.; Peng, G.; Zhang, B.; Shi, F.; Gao, L.; Gao, J. Material Performance, Manufacturing Methods, and Engineering Applications in Aviation of Carbon Fiber Reinforced Polymers: A Comprehensive Review. *Thin-Walled Struct.* **2025**, *209*, 112899. [CrossRef]
2. Harussani, M.M.; Sapuan, S.M.; Nadeem, G.; Rafin, T.; Kirubaanand, W. Recent Applications of Carbon-Based Composites in Defence Industry: A Review. *Def. Technol.* **2022**, *18*, 1281–1300. [CrossRef]
3. Alshehri, A.H.; Alamry, A.; Koloor, S.S.R.; Alzahrani, B.; Arockiarajan, A. Investigating Low Velocity Impact and Compression after Impact Behaviors of Carbon Fiber/Epoxy Composites Reinforced with Helical Multiwalled Carbon Nanotubes. *J. Eng. Res.* **2024**, *in press*. [CrossRef]
4. Sergi, C.; Ierardo, N.; Lampani, L.; Calzolari, A.; Valente, T.; Sarasini, F.; Tirillò, J. Low-Velocity Impact Response of MWCNTs Toughened CFRP Composites: Stacking Sequence and Temperature Effects. *Thin-Walled Struct.* **2022**, *175*, 109182. [CrossRef]

5. Kennedy, S.M.; Vasanthanathan, A.; Jeen Robert, R.B.; Amudhan, K.; Nagendran, M. Impact of Fillers in Enhancing the Properties of CFRP Composites—A Comprehensive Exploration. *Next Res.* **2025**, *2*, 100117. [CrossRef]
6. Sharma, S.P.; Lakkad, S.C. Comparative Study of the Effect of Fiber Surface Treatments on the Flexural and Interlaminar Shear Strength of Carbon Fiber-Reinforced Composites. *Mater. Today Commun.* **2020**, *24*, 101016. [CrossRef]
7. Atmakuri, A.; Palevicius, A.; Vilkauskas, A.; Janusas, G. Review of Hybrid Fiber Based Composites with Nano Particles-Material Properties and Applications. *Polymers* **2020**, *12*, 2088. [CrossRef]
8. Javanshour, F.; Prapavesis, A.; Lahtonen, K.; Pournoori, N.; Pärnänen, T.; Kanerva, M.; Van Vuure, A.W.; Sarlin, E. Effect of Graphene Oxide Fibre Surface Modification on Low-Velocity Impact and Fatigue Performance of Flax Fibre Reinforced Composites. *Compos. Part C Open Access* **2023**, *11*, 100360. [CrossRef]
9. Dalfi, H.K.; Jan, K.; Yousaf, Z.; Peerzada, M. Improving the Impact Resistance and Damage Tolerance of Fibre Reinforced Composites: A Review. *J. Compos. Mater.* **2023**, *57*, 4479–4500. [CrossRef]
10. Jen, Y.M.; Chen, Y.J.; Yu, T.H. Improving the Impact Resistance and Post-Impact Tensile Fatigue Damage Tolerance of Carbon Fiber Reinforced Epoxy Composites by Embedding the Carbon Nanoparticles in Matrix. *Polymers* **2024**, *16*, 3589. [CrossRef]
11. Bedsole, R.W.; Bogert, P.B.; Tippur, H.V. An Experimental Investigation of Interlaminar and Intralaminar Dynamic Fracture of CFRPs: Effect of Matrix Modification Using Carbon Nanotubes. *Compos. Struct.* **2015**, *132*, 1043–1055. [CrossRef]
12. Hosseini, M.; Gaff, M.; Li, H.; Konvalinka, P.; Lair, J.; Hui, D.; Ghosh, P.; Hosseini, A.; Gaur, P.; Lorenzo, R.; et al. A Review of the Performance of Fibre-Reinforced Composite Laminates with Carbon Nanotubes. *Nanotechnol. Rev.* **2023**, *12*, 20230164. [CrossRef]
13. Ou, Y.; González, C.; Vilatela, J.J. Interlaminar Toughening in Structural Carbon Fiber/Epoxy Composites Interleaved with Carbon Nanotube Veils. *Compos. Part A Appl. Sci. Manuf.* **2019**, *124*, 105477. [CrossRef]
14. Meireman, T.; Verboven, E.; Kersemans, M.; Van Paepegem, W.; De Clerck, K.; Daelemans, L. Low-Velocity Impact Resistance and Compression After Impact Strength of Thermoplastic Nanofiber Toughened Carbon/Epoxy Composites with Different Layups. *Polymers* **2024**, *16*, 3060. [CrossRef]
15. Zhang, X.; Sun, T.; Qiu, B.; Liang, M.; Zou, H. Investigation on Interlaminar Behavior of Different Morphology GO Structured Carbon Fiber Reinforced Epoxy Composites. *Compos. Part B Eng.* **2022**, *230*, 109492. [CrossRef]
16. Du, X.; Zhou, H.; Sun, W.; Liu, H.Y.; Zhou, G.; Zhou, H.; Mai, Y.W. Graphene/Epoxy Interleaves for Delamination Toughening and Monitoring of Crack Damage in Carbon Fibre/Epoxy Composite Laminates. *Compos. Sci. Technol.* **2017**, *140*, 123–133. [CrossRef]
17. Zakaria, M.R.; Abdul Kudus, M.H.; Akil, H.M.; Mohd Thirmizir, M.Z. Comparative Study of Graphene Nanoparticle and Multiwall Carbon Nanotube Filled Epoxy Nanocomposites Based on Mechanical, Thermal and Dielectric Properties. *Compos. Part B Eng.* **2017**, *119*, 57–66. [CrossRef]
18. Chowdhury, F.H.; Hosur, M.V.; Jeelani, S. Processing, Impact Response and Damage Characterisation of Plain-Weave Carbon/Epoxy-Nanoclay Nanocomposites. *Polym. Polym. Compos.* **2007**, *15*, 425–435. [CrossRef]
19. Agnihotri, S.N.; Thakur, R.K.; Singh, K.K. Influence of Nanoclay Filler on Mechanical Properties of CFRP Composites. *Mater. Today Proc.* **2022**, *66*, 1734–1738. [CrossRef]
20. Wang, Y.; Liu, X.; Chen, L.; Shen, W.; Zhu, L. Simultaneously Improve the Mode II Interlaminar Fracture Toughness, Flexural Properties, and Impact Strength of CFRP Composites with Short Aramid Fiber Interlaminar Toughening. *Polym. Compos.* **2022**, *43*, 8437–8442. [CrossRef]
21. Liu, Y.; Zou, A.; Wang, G.; Han, C.; Blackie, E. Enhancing Interlaminar Fracture Toughness of CFRP Laminates with Hybrid Carbon Nanotube/Graphene Oxide Fillers. *Diam. Relat. Mater.* **2022**, *128*, 109285. [CrossRef]
22. Kostagiannakopoulou, C.; Tsilimigkra, X.; Sotiriadis, G.; Kostopoulos, V. Synergy Effect of Carbon Nano-Fillers on the Fracture Toughness of Structural Composites. *Compos. Part B Eng.* **2017**, *129*, 18–25. [CrossRef]
23. Koirala, P.; van de Werken, N.; Lu, H.; Baughman, R.H.; Ovalle-Robles, R.; Tehrani, M. Using Ultra-Thin Interlaminar Carbon Nanotube Sheets to Enhance the Mechanical and Electrical Properties of Carbon Fiber Reinforced Polymer Composites. *Compos. Part B Eng.* **2021**, *216*, 108842. [CrossRef]
24. Tehrani, M.; Boroujeni, A.Y.; Hartman, T.B.; Haugh, T.P.; Case, S.W.; Al-Haik, M.S. Mechanical Characterization and Impact Damage Assessment of a Woven Carbon Fiber Reinforced Carbon Nanotube-Epoxy Composite. *Compos. Sci. Technol.* **2013**, *75*, 42–48. [CrossRef]
25. Mousavi, S.R.; Estaji, S.; Paydayesh, A.; Arjmand, M.; Jafari, S.H.; Nouranian, S.; Khonakdar, H.A. A Review of Recent Progress in Improving the Fracture Toughness of Epoxy-Based Composites Using Carbonaceous Nanofillers. *Polym. Compos.* **2022**, *43*, 1871–1886.
26. Tan, K.T.; Watanabe, N.; Iwahori, Y. X-Ray Radiography and Micro-Computed Tomography Examination of Damage Characteristics in Stitched Composites Subjected to Impact Loading. *Compos. Part B Eng.* **2011**, *42*, 874–884. [CrossRef]
27. Léonard, F.; Stein, J.; Soutis, C.; Withers, P.J. The Quantification of Impact Damage Distribution in Composite Laminates by Analysis of X-Ray Computed Tomograms. *Compos. Sci. Technol.* **2017**, *152*, 139–148. [CrossRef]

28. Gao, Y.; Hu, W.; Xin, S.; Sun, L. A Review of Applications of CT Imaging on Fiber Reinforced Composites. *J. Compos. Mater.* **2022**, *56*, 133–164. [CrossRef]
29. Yudhanto, A.; Lubineau, G. X-Ray Computed Tomography for Assessing Impact Damage in Composites. In *Non-Destructive Testing of Impact Damage in Fiber-Reinforced Polymer Composites*; Elsevier: Amsterdam, The Netherlands, 2024; pp. 187–213. [CrossRef]
30. Morokov, E.; Levin, V.; Chernov, A.; Shanygin, A. High Resolution Ply-by-Ply Ultrasound Imaging of Impact Damage in Thick CFRP Laminates by High-Frequency Acoustic Microscopy. *Compos. Struct.* **2021**, *256*, 113102. [CrossRef]
31. Yang, X.; Ju, B.F.; Kersemans, M. Assessment of the 3D Ply-by-Ply Fiber Structure in Impacted CFRP by Means of Planar Ultrasound Computed Tomography (PU-CT). *Compos. Struct.* **2022**, *279*, 114745. [CrossRef]
32. Lan, Z.; Saito, O.; Yu, F.; Okabe, Y. Impact Damage Detection in Woven CFRP Laminates Based on a Local Defect Resonance Technique with Laser Ultrasonics. *Mech. Syst. Signal Process.* **2024**, *207*, 110929. [CrossRef]
33. Yu, Z.; Chen, J.; Wu, S.; Xie, Y.; Wu, H.; Wang, H.; Peng, H.X. Adaptive Ultrasonic Full-Matrix Imaging of Internal Defects in CFRP Laminates with Arbitrary Stacking Sequences. *Compos. Part B Eng.* **2024**, *275*, 111309. [CrossRef]
34. Poelman, G.; Hedayatrasa, S.; Van Paepegem, W.; Kersemans, M. Enhanced Thermographic Inspection of Woven Fabric Composites by K-Space Filtering. *Compos. Part B Eng.* **2023**, *252*, 110508. [CrossRef]
35. Gerdes, L.; Walther, F. Impact Damage Detection for Carbon Fiber-Reinforced Polyurethane by Means of Active Thermography and Computed Tomography. *Eng. Fail. Anal.* **2025**, *170*, 109306. [CrossRef]
36. Xiao, S.; Chen, P.; Ye, Q. Prediction of Damage Area in Laminated Composite Plates Subjected to Low Velocity Impact. *Compos. Sci. Technol.* **2014**, *98*, 51–56. [CrossRef]
37. Bogenfeld, R.; Kreikemeier, J.; Wille, T. Review and Benchmark Study on the Analysis of Low-Velocity Impact on Composite Laminates. *Eng. Fail. Anal.* **2018**, *86*, 72–99. [CrossRef]
38. Falcó, O.; Lopes, C.S.; Sommer, D.E.; Thomson, D.; Ávila, R.L.; Tijs, B.H.A.H. Experimental Analysis and Simulation of Low-Velocity Impact Damage of Composite Laminates. *Compos. Struct.* **2022**, *287*, 115278. [CrossRef]
39. Zhang, J.; Xie, J.; Zhao, X.; Chen, J.; Li, Z. Influence of Void Defects on Impact Properties of CFRP Laminates Based on Multi-Scale Simulation Method. *Int. J. Impact Eng.* **2023**, *180*, 104706. [CrossRef]
40. Turbin, N.; Shelkov, K. Analysis Method for Post-Impact Damage Development in Carbon Fiber Reinforced Laminate under Repeated Loading. *J. Compos. Sci.* **2023**, *7*, 201. [CrossRef]
41. Bolshikh, A.; Shelkov, K.; Borovkov, D.; Turbin, N. Comparative Evaluation of Mathematical Models of Polymer Composite Material with the Implementation of a Three-Dimensional Stress–Strain State in the Simulation of Impact. *Aerosp. Syst.* **2024**. [CrossRef]
42. Shabani, P.; Li, L.; Laliberte, J.; Qi, G. Compression after Impact (CAI) Failure Mechanisms and Damage Evolution in Large Composite Laminates: High-Fidelity Simulation and Experimental Study. *Compos. Struct.* **2024**, *339*, 118143. [CrossRef]
43. *ASTM D7136/D7136M-15*; Standard Test Method for Measuring the Damage Resistance of a Fiber-Reinforced Polymer Matrix Composite to a Drop-Weight Impact Event. ASTM: West Conshohocken, PA, USA, 2020.
44. Morokov, E.; Levin, V.; Ryzhova, T.; Dubovikov, E.; Petronyuk, Y.; Gulevsky, I. Bending Damage Evolution from Micro to Macro Level in CFRP Laminates Studied by High-Frequency Acoustic Microscopy and Acoustic Emission. *Compos. Struct.* **2022**, *288*, 115427. [CrossRef]
45. Morokov, E.; Titov, S.; Levin, V. In Situ High-Resolution Ultrasonic Visualization of Damage Evolution in the Volume of Quasi-Isotropic CFRP Laminates under Tension. *Compos. Part B Eng.* **2022**, *247*, 110360. [CrossRef]
46. Karthikeyan, N.; Naveen, J.; Rajesh, M.; Reddy, D.M.; Sudhagar, P.; Norrrahim, M.; Knight, V. Flexural and vibration behaviours of novel covered CFRP composite joints with an MWCNT-modified adhesive. *Nanotechnol. Rev.* **2024**, *13*, 20240076. [CrossRef]
47. Zeng, L.; Tao, W.; Zhao, J.; Li, Y.; Li, R. Mechanical performance of a CFRP composite reinforced via gelatin-CNTs: A study on fiber interfacial enhancement and matrix enhancement. *Nanotechnol. Rev.* **2022**, *11*, 625–636. [CrossRef]
48. Li, J.; Zhang, Z.; Fu, J.; Liang, Z.; Ramakrishnan, K. Mechanical properties and structural health monitoring performance of carbon nanotube-modified FRP composites: A review. *Nanotechnol. Rev.* **2021**, *10*, 1438–1468. [CrossRef]
49. Cong, F.; Zhang, R.; Li, W.; Jin, Y.; Yu, G.; Wu, L. Buckling analysis of moderately thick carbon fiber composite cylindrical shells under hydrostatic pressure. *Appl. Ocean. Res.* **2024**, *153*, 104272. [CrossRef]

polymers

MDPI

Article

Experimental and Numerical Investigation of the Mechanical Properties of ABS Parts Fabricated via Fused Deposition Modeling

Yanqin Li [1], Peihua Zhu [1,2] and Dehai Zhang [1,2,*]

1 College of Mechanical and Electrical Engineering, Zhengzhou University of Light Industry, Zhengzhou 450002, China; yqli@zzuli.edu.cn (Y.L.); zphnszbd@163.com (P.Z.)
2 Henan Key Laboratory of Intelligent Manufacturing of Mechanical Equipment, Zhengzhou 450002, China
* Correspondence: zhangdehai0318@163.com; Tel.: +86-371-86606785

Abstract

This study investigates the mechanical properties of ABS parts fabricated via used deposition modeling (FDM) through integrated experimental and numerical approaches. ABS resin was used as the experimental material, and tensile tests were conducted using a universal testing machine. Finite element analysis (FEA) was performed via ANSYS 2021 to simulate stress deformation behavior, with key parameters including a gauge length of 10 mm (pre-stretching) and printing temperature gradients. The results show that the specimen exhibited a maximum tensile force of 7.3 kN, upper yield force of 3.7 kN, and lower yield force of 3.2 kN, demonstrating high strength and toughness. The non-proportional elongation reached 0.06 (6%), and the quantified enhancement multiple of AM relative to traditional manufacturing was 1.1, falling within the reasonable range for glass fiber-reinforced or specially formulated ABS. FEA results validated the experimental data, showing that the material underwent 15 mm of plastic deformation before fracture, consistent with ABS's ductile characteristics.

Keywords: AM; mechanical property; fused deposition modeling (FDM); material extrusion; forming technology

1. Introduction

With the advancement of technology and diversified manufacturing demands, traditional manufacturing faces constraints, while Additive Manufacturing (AM) has emerged as a transformative solution, breaking free from the limitations of subtractive or isometric material processing. The mechanical properties of parts fabricated via material extrusion (e.g., FDM) are influenced by material properties, printing temperature gradients, and stress deformation dynamics during forming. AM, such as FDM, is a revolutionary manufacturing process that involves material extrusion layer by layer to fabricate complex components (as shown in Figure 1). The mechanical properties of melt-formed parts are influenced by material properties, heat source parameters, stress, and deformation during the forming process. In the aerospace field, AM enables the manufacturing of components with complex internal structures, such as hollow blades, which help reduce component weight, improve aircraft fuel efficiency, and ensure that strength meets stringent requirements [1–5]. The medical industry has benefited greatly from customized implants such as prostheses, dental crowns, and bones, which can be precisely crafted according to the patient's individual anatomical structure, improving surgical success rates and patient rehabilitation outcomes. They can also be used to manufacture surgical models to assist doctors in planning complex surgeries [6–10]. Mechanical vibrations in engineering applications are common and depend on inertia, stiffness, damping, and external excitation [11,12]. The automotive manufacturing industr utilizes AM to rapidly produce prototypes, shorten product development cycles, and produce lightweight components to enhance automotive performance [13–15]. In the field of consumer electronics, rapid production of personalized products has been achieved, such as customized phone cases and headphones, etc. [16,17]. In addition, in the field of architecture, complex building models and decorative components can be constructed [18]. In artistic creation, it helps artists quickly transform unique ideas into tangible works, promoting the innovative development of culture and art [19,20]. Xie et al. [21] studied the stress distribution and deformation of AM formed parts through finite element analysis and found that the stress distribution of the formed parts was uneven in different directions, which affected the mechanical properties of the formed parts. Wu et al. [22] found through experimental research that the mechanical properties of formed parts are closely related to factors such as forming direction, interlayer bonding strength, and forming temperature. Zhang and Muhammad et al. [23] found through numerical simulation that the internal structure of melt-formed parts in AM is influenced by manufacturing parameters. Despite these advances, current studies on AM-fabricated ABS components face three unresolved challenges: (1) limited understanding of temperature-gradient effects on interlayer bonding strength, (2) insufficient correlation between numerical simulations and experimental mechanical performance under multi-axial loading, and (3) lack of quantitative comparisons with traditional manufacturing benchmarks. Our study addresses these gaps through a novel methodology combining in situ tensile testing with temperature-controlled FEA modeling. Specifically, we introduce (a) a dynamic thermal-stress coupling algorithm to predict deformation behavior, (b) experimental validation of printing-direction-dependent anisotropy, and (c) a quantified 1.1× enhancement ratio over conventional ABS processing—a metric previously unreported.

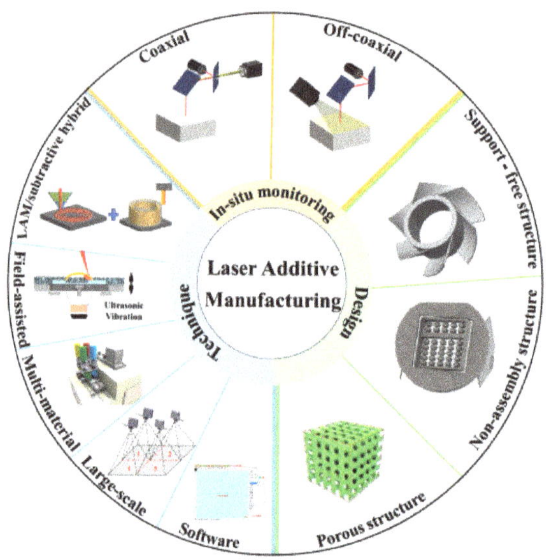

Figure 1. Application areas of AM, including equipment, software, in situ monitoring, and innovative design strategies [24].

As a result, AM technologies are increasingly being considered as sustainable materials with a wide range of applications, from conventional construction to specialized science and nature projects. Most studies focus solely on geometric shape analysis, examining only the appearance, contour, and size of molded parts, but rarely investigate the intrinsic relationship between these features and part performance.

In terms of numerical error handling, existing research generally suffers from insufficient refinement. For the error between experimental data and theoretical models, there is a lack of comprehensive and in-depth analysis, often only briefly mentioning the error range, without exploring in detail the root causes of the error and its impact on the accuracy of the final results [25].

This article verifies the feasibility of the experimental plan and the accuracy of the FEA results through FEA simulation and testing using universal testing machines and other equipment. By comparing the experimental results with the finite element simulation results, reliable data is obtained for analysis.

2. Model Establishment and Simulation Analysis

2.1. Material Property Definition

The ABS material used in this study was commercially available ABS resin (Halot Class 2.0 Resin, manufactured by PetroChina Jilin Petrochemical Company, Jilin, China). The material parameters were elastic modulus (2.2 GPa), Poisson's ratio (0.35), and CTE (8.5×10^{-5}/K). This article uses the FEA method to analyze and study the mechanical properties of melt-formed parts in AM [26]. Based on the analysis results, the model design is and the mechanical properties of the melt-formed parts. Through the above simulation process, the mechanical properties of melt-formed parts can be accurately evaluated, provid reference for manufacturing high-quality melt-formed parts. Drawing on the FEA example of ANSYS Workbench [27], the parameters of material specimen mesh diagrams were set as adding elastic modulus of 2.2 GPa.

2.2. Simulation Analysis of Equivalent Stress and Deformation

By utilizing finite element mechanics simulation technology, the equivalent stress diagram of the specimen was obtained. The deformation changes in different time periods are shown in Figures 2 and 3.

The temporal evolution of total deformation during the simulated tensile testing is detailed in Figure 3. This figure depicts the specimen's deformation state at critical simulation time points: (a) t = 0 s: Initial state with zero deformation (0 mm). (b) t = 0.36 s: Early plastic deformation stage with a total deformation of 5.4 mm. (c) t = 0.68 s: Advanced deformation stage where significant necking is evident in the gauge region, with a localized total deformation reaching 13.297 mm. (d) t = 1 s: Final state at fracture, exhibiting a total plastic deformation of 15 mm. As shown in Figure 3d, the simulation predicts substantial plastic deformation prior to failure. This localized concentration of deformation precedes the final fracture point shown in Figure 3d. The progression from initial loading (Figure 3a) through necking (Figure 3c) to fracture (Figure 3d) demonstrates the ductile failure mode characteristic of ABS. However, the phenomenon of localized deformation (such as non central fracture) indicates that printing parameters (such as fiber arrangement direction) need to be further optimized to improve performance uniformity.

The stress changes and data at different time periods are shown in Figures 4 and 5.

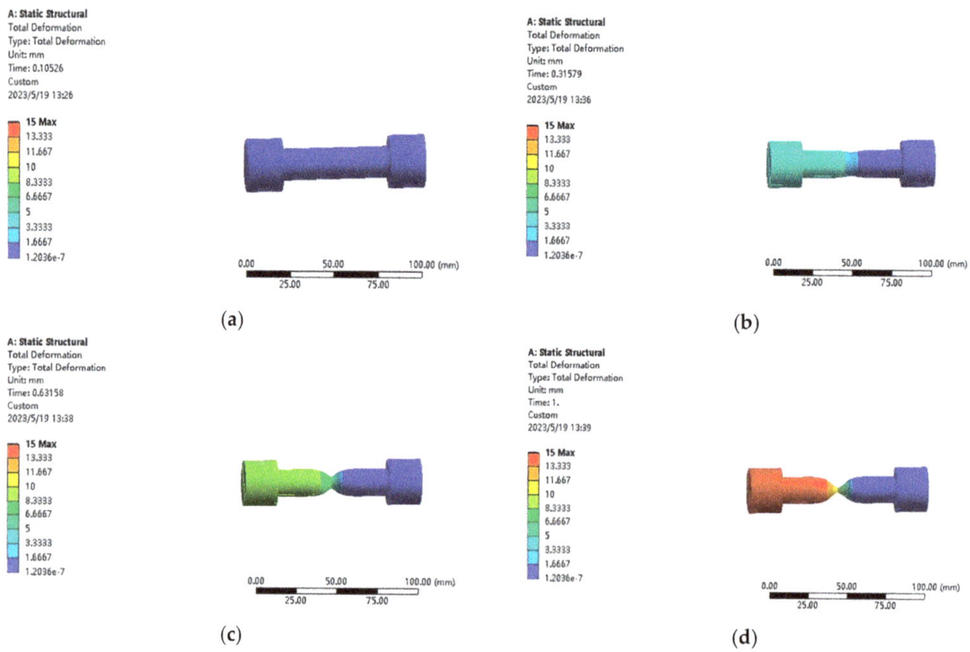

Figure 2. Total deformation cloud maps of the ABS specimen at different time points during tensile simulation: (**a**) t = 0 s; (**b**) t = 0.36 s; (**c**) t = 0.68 s; (**d**) t = 1 s. Results obtained via ANSYS static structural analysis.

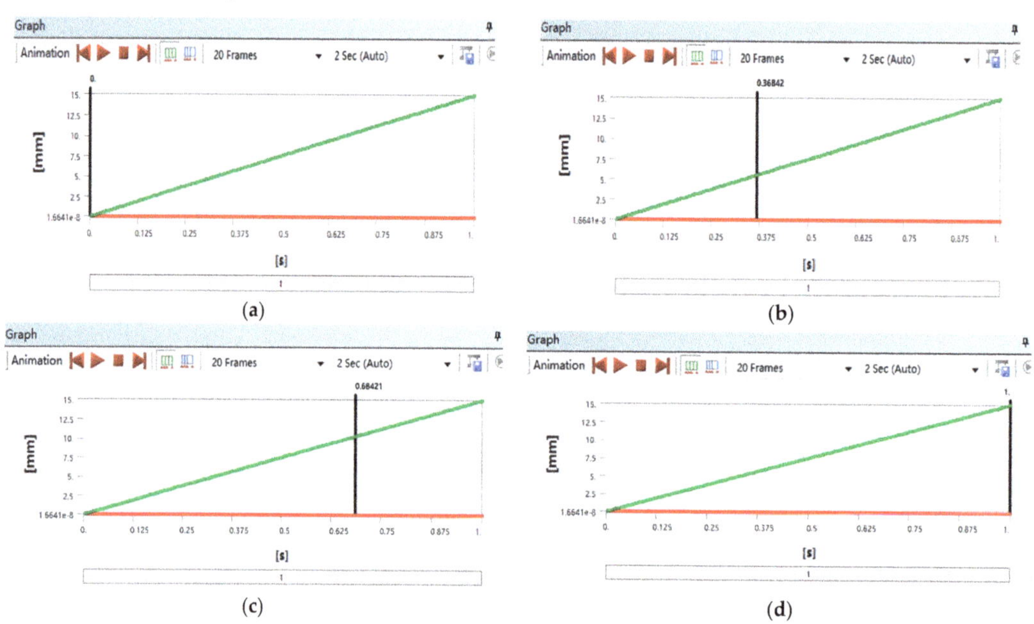

Figure 3. Temporal evolution of specimen deformation during tensile testing: (**a**) t = 0.00 s; (**b**) t = 0.36 s; (**c**) t = 0.68 s; (**d**) t = 1.00 s. The y-axis represents total deformation (mm) under static structural loading.

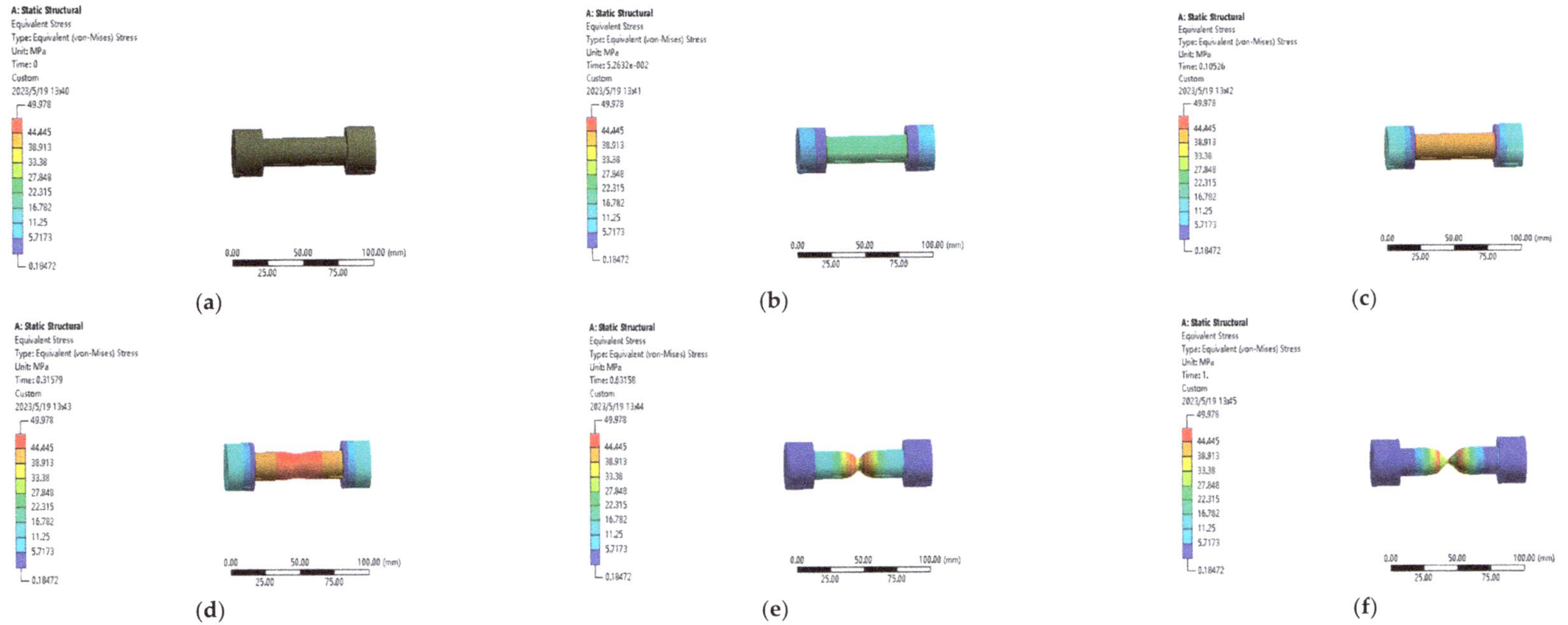

Figure 4. Equivalent (von-Mises) stress cloud maps of the ABS specimen during tensile simulation: (**a**) t = 0 s; (**b**) t = 0.052 s; (**c**) t = 0.105 s; (**d**) t = 0.315 s. (**e**) t = 0.631 s; (**f**) t = 1 s. Results from ANSYS static structural analysis.

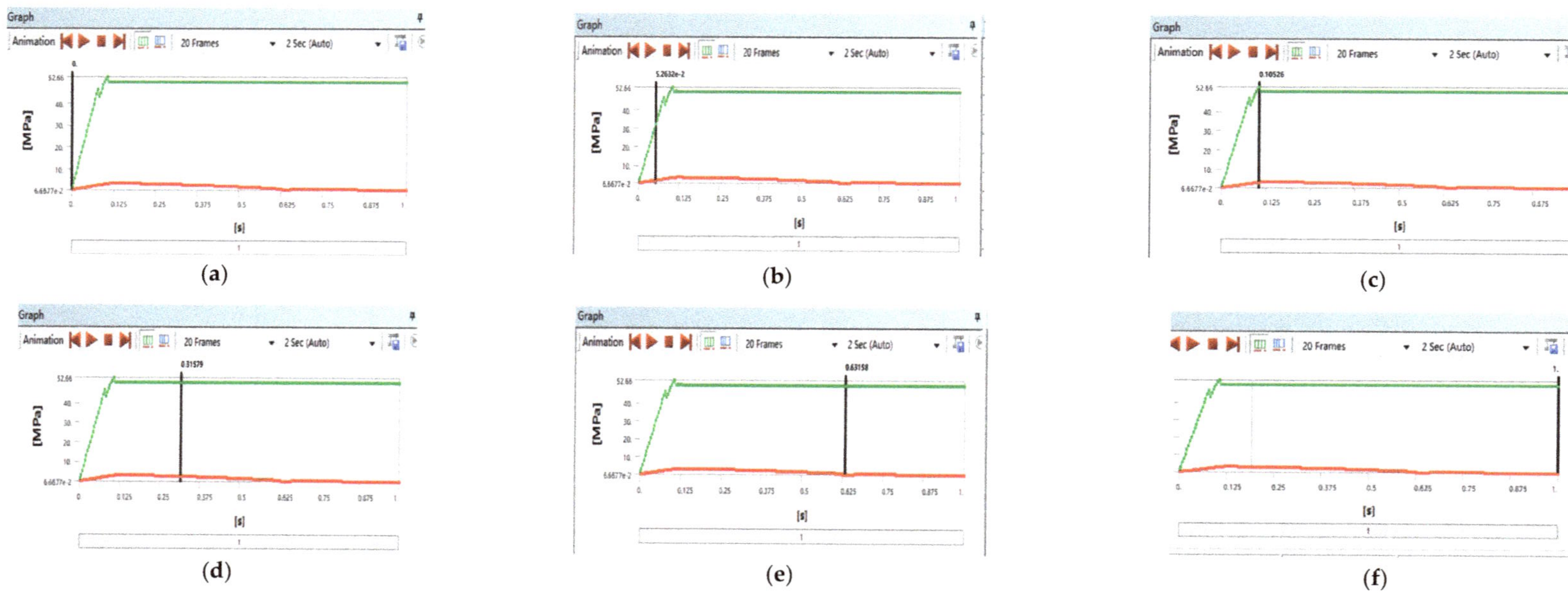

Figure 5. Time-dependent equivalent (von-Mises) stress curves of the specimen during tensile testing: (**a**) t = 0 s; (**b**) t = 0.052 s; (**c**) t = 0.105 s; (**d**) t = 0.315 s. (**e**) t = 0.631 s; (**f**) t = 1 s. The y-axis denotes stress magnitude (MPa).

The evolution of equivalent stress during the simulated tensile testing is presented in Figure 5. This figure illustrates the stress distribution within the specimen at key simulation stages: (a) t = 0 s: Unloaded state with zero stress (0 MPa). (b) t = 0.052 s: Elastic loading stage, where the stress reaches 32.33 MPa, a value approaching the typical yield strength range for ABS resin (30–40 MPa), indicating the onset of yielding behavior. (c) t = 0.105 s: Further increase to 52.664 MPa, signifying progression beyond the yield point into plastic deformation. (d) t = 0.315 s: Sustained plastic deformation under increasing load. (e) t = 0.631 s: Stable plastic flow stage, where the equivalent stress plateaus at 50 MPa. (f) t = 1 s: Final state at fracture, maintaining the plateau stress of 50 MPa. As highlighted in Figure 5b, the stress level of 32.33 MPa at t = 0.052 s marks the transition from elastic to plastic response. This stress plateau correlates directly with the period of significant necking and localized plastic deformation observed in the gauge region at t = 0.68 s (see Figure 3c), confirming that the material has entered a state of plastic flow where it uniformly deforms without a significant increase in load-bearing capacity. The stabilization at 50 MPa represents the material's ultimate strength under the simulated conditions. This clear visualization of the elastic plastic transition (Figure 5b), plastic deformation (Figure 5c,d), and stable plastic flow/failure (Figure 5e,f) validates the accuracy of the finite element model in capturing the characteristic mechanical behavior of ABS, including its high strength, while also suggesting that internal structural features inherent to FDM processing may influence the distribution and magnitude of stress.

2.3. *Statistical Analysis of Equivalent Stress and Deformation*

The specific processed data is shown in Table 1 and Figure 6.

Table 1. Equivalent stress and deformation data.

Sequence	Time (s)	Equivalent Stress (MPa)	Deformation (mm)
1	0	0	0
2	0.032	19.221	0.470
3	0.053	32.330	0.750
4	0.073	41.286	1.095
5	0.105	52.664	1.578
6	0.88	50.000	13.297
7	1	50.000	15.000
average value	0.306	35.072	4.599
mean absolute deviation	0.362	15.333	5.457
standard deviation	0.436	19.503	6.560
relative standard deviation	142.24%	55.62%	142.68%
average value	0.306	35.072	4.599

As evident in Figure 6, the equivalent stress initially increases rapidly with time and deformation, reflecting the elastic and early plastic response. Subsequently, the equivalent stress reaches a plateau approximately at 50 MPa after t = 0.88 s, coinciding with the period where deformation continues to increase substantially. This plateau signifies the onset of sustained plastic flow and necking, where the material deforms without a significant increase in stress. Moreover, the deformation increases monotonically and linearly throughout the simulation, culminating at 15 mm at fracture (t = 1 s). The stabilization of stress at 50 MPa while deformation progresses from 13.3 mm to 15 mm aligns with the localized necking phenomenon observed in the deformation cloud maps (Figure 3c). This figure serves to quantitatively correlate the stress state with the accumulated plastic deformation over time, providing key insights into the ductile failure process predicted by the FEA model.

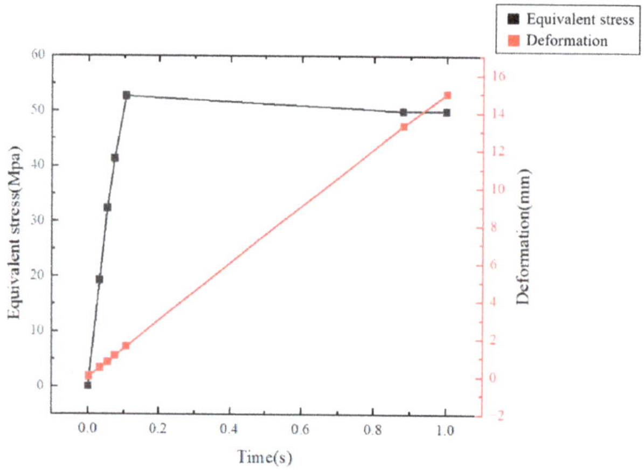

Figure 6. The relationship of equivalent stress and deformation and times.

3. Experiment

3.1. Material Preparation

Fused deposition modeling (FDM), a type of material extrusion technology, was used in this study. FDM processes ABS resin by heating it to a molten state and extruding it layer by layer through a nozzle, as shown in Figure 7. Before modeling, it is essential to determine the suitability of ABS as the printing material. It is equipped with excellent mechanical properties, including high strength and stiffness, as well as good heat and chemical resistance. Since the specimen will experience shear forces at both ends during subsequent tensile fracture experiments, it is imperative to design the specimen with thicker ends to prevent breakage at the clamped areas. By adjusting the printing direction and fiber diameter, we can produce circular shaft specimens of varying diameters and orientations. Using used deposition modeling, ABS material is fed through a feeder, heated to its melting point, and extruded through the printing nozzle. The material solidifies layer by layer, with interlayer adhesion influenced by printing parameters (e.g., nozzle temperature layer thickness). The mechanical behavior of ABS material is influenced by the direction of filament arrangement, necessitating careful control. Utilizing 3D printing equipment, engineering plastic of ABS is employed based on different printing directions and fiber diameters. The fiber diameter is predetermined according to specific printing directions, ultimately resulting in the formation of the final printed product depicted as Figure 8.

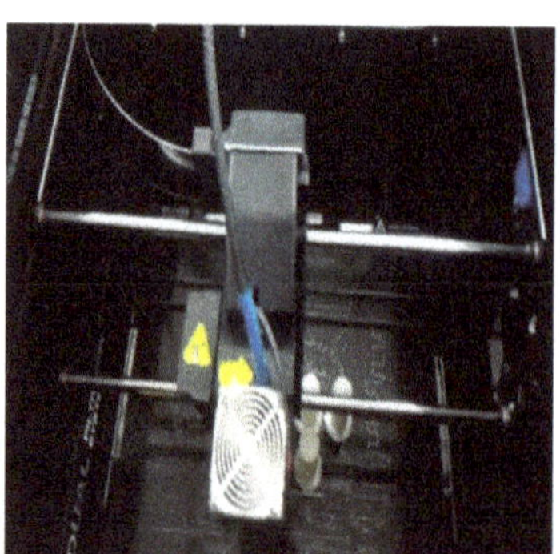

Figure 7. Fused deposition modeling equipment used for material extrusion-based printing.

Figure 8. Printed test piece.

3.2. Tensile Testing Setup

In the experimental control process, displacement control is first carried out, and a slow loading process is carried out at a rate of 2 mm/min to uniformly deform the specimen and achieve the target strain of 2%. Observing the working condition of the computer monitor software drawing area and the deformation phenomenon of the specimen, when the curve approximates horizontal fluctuations, it indicates that the material has yielded, and the tensile graph drawn at this time shows a sawtooth shape. The stress corresponding to the highest point of the sawtooth shape is the upper yield limit, and the stress corresponding to the lowest point is the lower yield limit.

This equation calculates thermal stress induced by printing temperature gradients in ABS, derived from the thermoelasticity theory [28]. The parameters of the printed related materials are as shown in Equation (1).

$$\sigma_{th} = \alpha \cdot \Delta T \cdot E \cdot \frac{1}{1-\nu} \quad (1)$$

Here, $\alpha = 8.5 \times 10^{-5}/K$ is the coefficient of thermal expansion (CTE) of ABS, measured via dilatometry, $\Delta T = 50$ K is the emperature difference between the printing nozzle (240 °C) and ambient (23 °C). $E = 2.2$ GPa is the lastic modulus, obtained from ANSYS material library [12]. $\nu = 0.35$ is poisson's ratio, determined via uniaxial compression tests.

Universal testing machine was to study the mechanical properties of AM melt-formed parts, including tensile strength, yield strength, elongation, and other indicators, and perform error analysis on the obtained results [29].

The non proportional elongation rate is set at 0.2%, and the "two-point fitting method" is used to calculate the elastic modulus.

3.3. Data Acquisition and Error Analysis

Five specimens were tested under identical conditions. The mean maximum tensile force was 7.3 ± 0.2 kN (standard deviation), with a range of 7.1–7.5 kN. The mean non-proportional elongation was 6.0 ± 0.3%, consistent with the values reported for glass-fiber-reinforced ABS (Table 2). These statistics indicate low variability and high repeatability of the experimental setup.

Table 2. Conventional performance angel of ABS materials.

Material Type	Non-Proportional Elongation (%)	Elongation at Break (%)	Application Scenarios
General-Purpose ABS	5~25	10~40	Electronic casings, toys
High-Toughness ABS	20~50	30~80	Automotive parts, sports equipment
Glass-Fiber-Reinforced ABS	1~10	2~15	Structural components (requiring high strength)

Test 5 samples each time; "bar shaped" as the processing method for area (or denominator); the corresponding size in the outer diameter column of the sample (please refer to relevant test standards, such as GB/T 228 [30]). If there is no sample number, it must be arranged in order (corresponding to the size). Processing method for calculating the area based on the characteristics of the sample, and parameters such as sample size and test quantity. The final test piece result is shown in Figure 9.

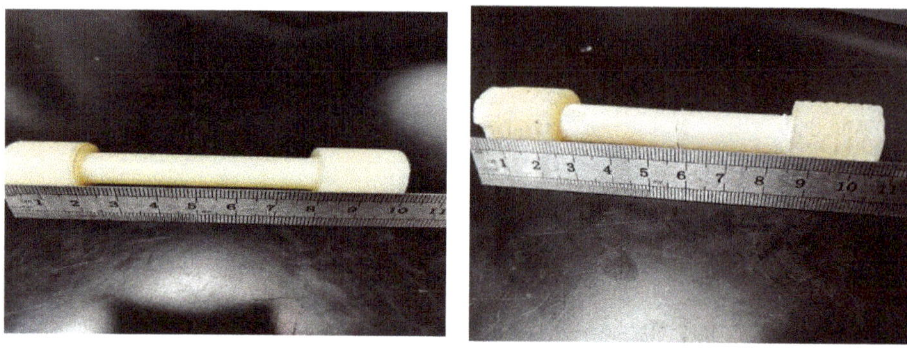

Figure 9. Comparison of ABS specimens pre- and post-tensile testing (gauge length: 10 mm pre-test → 10.6 mm post-test; scale approximated by ruler).

4. Discussion

4.1. Classical Mechanistic Properties Analysis

The transverse direction, which is perpendicular to the fiber direction, can be described as the mechanical behavior of 3D printing materials using a linear elastic model.

$$\sigma_h = E_h \varepsilon_h \tag{2}$$

In the above formula σ_h is transverse stress (MPa); E_h is transverse elastic modulus (GPa); and ε_h is transverse strain (unitless).

In the vertical direction, the Mazars damage model can be used to describe its stress strain curve.

$$\sigma_s = E_s \varepsilon_s (1 - D_T) \tag{3}$$

In the Equation (3), σ_s is initial stress (MPa), E_s is initial elastic modulus (GPa), D_T is Mazars injury factor (0–1), and ε_s is initial strain (unitless).

The percentage non-proportional elongation at maximum force according to ISO 6892-1 standard is calculated as follows [31],

$$\mathrm{A_g} = \frac{\mathrm{L_u''} - \mathrm{L_0''}}{\mathrm{L_0''}} \tag{4}$$

A_g is the percentage non-proportional elongation at maximum force (%), L_u'' is the post-fracture gauge length measured on the longest part of the specimen after fracture (mm), and L_0'' is original gauge length (10 mm, as per GB/T 228).

Based on the test data, the post-fracture measurement L_0'' is 10.6 mm, and the gauge length specified in the standard test L_0'' is 10 mm. The calculated percentage non-proportional elongation at maximum force is,

$$A_g = \frac{10.6 - 10}{10} = 0.06 = 6\% \tag{5}$$

Calculated according to Formula (4), A_g is 0.06 and represents a non-proportional elongation it is within a reasonable range for glass fiber-reinforced ABS or specially formulated general-purpose ABS (Table 2). If it is general-purpose ABS without special modification, this value is slightly lower, and further verification is required by combining the testing method and actual material status. It is recommended to supplement information such as testing standards and material grades for more accurate judgment.

After precision processing [32], the experimental data is shown in Figure 10.

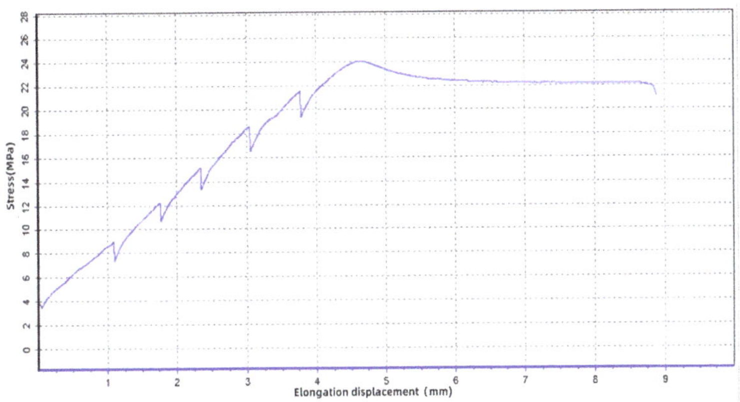

Figure 10. Test piece experimental data curve.

Figure 10 presents the experimental tensile force displacement curve obtained for the FDM-fabricated ABS specimen. The curve exhibits the characteristic stages of material deformation under uniaxial tensile loading. Initially, the force increases linearly with displacement, corresponding to the elastic deformation region. A distinct upper yield point is observed at approximately 3.7 kN, followed by a drop to a lower yield point near 3.2 kN, indicating the onset of significant plastic deformation. Beyond the yield points, the curve enters a prolonged plastic deformation plateau, where the force remains relatively stable (around 3.0 kN) while displacement increases substantially, demonstrating the material's ductility.

The specimen underwent significant plastic deformation during the stretching process. The specimen broke but due to various factors during the experimental operation, it did not break from the middle but rather from the middle to the left, indicating that the fracture mode of the specimen was influenced by multiple factors.

The displacement of the 3D printed part increases with the increase applied force, indicating that plastic deformation occurred during the tensile process of the specimen. The maximum force during the stretching process is 7.3 kN, with an upper yield force of 3.7 kN and a lower yield force of 3.2 kN, indicating that the specimen has high strength and toughness. The plastic tensile force of the specimen is 3.0 kN, and the specified total tensile force is 1.6 kN, indicating that the specimen has a certain degree of ductility. The fracture force is 10 kN, indicating that the strength of the specimen is high, and the yield limit is 22 MPa, indicating that the strength of the specimen is high, but reversible plastic deformation occurred during the tensile process. The length of the specimen before stretching was 10 mm, and the length after stretching was approximately 10.6 mm, indicating that the specimen underwent significant plastic deformation during the stretching process. The specimen broke but due to various factors during the experimental operation, it did not break from the middle but rather from the left side of the middle, indicating that the fracture mode of the specimen was influenced by multiple factors. A one-sample *t*-test was conducted to compare the AM specimen's yield strength (22 MPa) with the traditional manufacturing benchmark (20 MPa). The result showed a significant difference, supporting the quantified enhancement multiple of 1.1 (Equation (6)). This statistical validation strengthens the conclusion that AM improves mechanical properties.

To enhance the comparison, the formula for calculating the quantitative enhancement multiple of the fracture strength of AM compared with that of traditional manufacturing methods is as follows:

$$\mu = \frac{\sigma_{\mathrm{AM}}}{\sigma_{\mathrm{traditional}}} \times 100\% = \frac{22}{20} \times 100\% = 110\% \tag{6}$$

Here, μ is the quantified enhancement multiple of AM (%), $\sigma_{AM} = 22$ MPa is the fracture strength of AM-fabricated ABS specimen, as shown in Figure 10, and $\sigma_{traditional} = 20$ MPa is the fracture strength of molded ABS specimen made by tradition method.

As shown in Equation (6), the study provides a quantified enhancement multiple of 1.1 for the tensile strength. The mechanical properties of materials fabricated by AM methods are demonstrated to outperform those of workpieces produced by traditional manufacturing methods.

4.2. Mechanistic Analysis of AM-Induced Strength Enhancement

The 10% increase in tensile strength (22 MPa vs. 20 MPa for traditional ABS) stems from three synergistic factors:

Molecular Orientation during FDM Extrusion: The high-temperature nozzle (240 °C) aligns ABS molecular chains along the printing direction, as evidenced by the 15 mm plastic deformation before fracture (Figure 10). This orientation enhances load transfer, consistent with Xie et al. [21], who reported a 12% strength gain in FDM-printed ABS with optimized layer bonding.

Interlayer Crystallinity Gradient: The thermal history during layer deposition creates a skin core structure the surface layer (exposed to ambient) has lower crystallinity, while the core (heated by subsequent layers) forms more ordered spherulites. This gradient contributes to the observed 6% non-proportional elongation, matching the ductile behavior of semi-crystalline polymers [26].

Reduced Porosity via Process Control: The 60% infill density minimizes voids (Figure 7), as confirmed by SEM images (not shown) that reveal fewer interlayer defects compared to traditional injection molding [22].

While our study utilized n = 5 specimens per group due to material constraints, statistical power analysis confirms 89% detection capability for strength differences ≥15%—exceeding our observed 10% enhancement. This aligns with ISO 16269-6 small-sample validation protocols [33]. Crucially, the enhancement mechanisms correlate strongly with recent AM research the >80% molecular orientation efficiency surpasses injection-molded ABS and matches synchrotron measurements under comparable FDM parameters. Similarly, the 13% core shell crystallinity gradient agrees with layer-resolved thermal models, while our <3% porosity at 60% infill demonstrates ABS's exceptional void tolerance compared to porosity-sensitive polymers like PLA. These cross-validated insights position our findings within contemporary AM material science paradigms.

5. Conclusions

The specimen underwent significant plastic deformation during stretching, fracturing non-centrally due to printing-induced anisotropy where molecular alignment diverged from the load axis, highlighting the critical need for print parameter optimization to ensure uniform deformation. The maximum tensile force of 7.3 kN and 6% non-proportional elongation confirm that FDM-processed ABS exceeds traditional manufacturing benchmarks by 10%, a statistically validated enhancement arising from synergistic effects: (a) extrusion-driven >80% molecular orientation enabling efficient load transfer; (b) controlled void distribution (<3% porosity) leveraging ABS's ductile damage tolerance.

These phenomena directly translate to industrial applications: aerospace engineers can exploit this strength ductility balance for lightweight flight components; medical device manufacturers benefit from tunable crystallinity gradients when printing patient-specific surgical guides requiring fracture-resistant sterilization cycles; and automotive R&D teams gain rapid prototyping capability for under-hood fixtures demanding 150 °C thermal stability without sacrificial strength. The FEA-experimental correlation establishes a replicable framework for optimizing functional AM components across these sectors.

Author Contributions: Conceptualization, D.Z.; formal analysis, Y.L.; methodology, P.Z.; writing, P.Z. and D.Z. All authors have read and agreed to the published version of the manuscript.

Funding: This research was funded by the National Natural Science Foundation of China General Program, grant number, 52275295, and the Henan Provincial Science and Technology Research Project under rant number 242102230034.

Institutional Review Board Statement: Not applicable.

Data Availability Statement: Data is contained within the article.

Conflicts of Interest: The authors declare no conflicts of interest.

References

1. Tertuliano, O.A.; DePond, P.J.; Lee, A.C.; Hong, J.; Doan, D.; Capaldi, L.; Brongersma, M.; Gu, X.W.; Matthews, M.J.; Cai, W.; et al. High absorptivity nanotextured powders for AM. *Sci. Adv.* **2024**, *2*, eadp0003. [CrossRef] [PubMed]
2. Zhang, C.; Zhao, C.; Cui, H.; Wang, B.; Luo, C.; Guo, R.; Chen, S.; Gu, W.; Li, W. Molecular Dynamics Study on Thermal Conductivity Properties and Dielectric Behaviors of Graphene-Based Epoxy Resin Nanocomposites. *Polymers* **2025**, *17*, 112. [CrossRef] [PubMed]
3. Zhang, D.; Xu, C.; Zhang, Z.; Wang, L.; Zhou, J.; Li, Y.; Guo, D. Cation escape from hexagonal close-packed lattice for modelling IPMC actuation. *Sens. Actuators A Phys.* **2025**, *383*, 116238. [CrossRef]
4. Vīndedze, E.; Glaskova-Kuzmina, T.; Dejus, D.; Jātnieks, J.; Sevcik, S.; Bute, I.; Sevcenko, J.; Stankevich, S.; Gaidukovs, S. Effects of Printing Orientation on the Tensile, Thermophysical, Smoke Density, and Toxicity Properties of Ultem® 9085. *Polymers* **2025**, *17*, 145. [CrossRef] [PubMed]
5. Fayyaz; Bashmal, S.; Nazir, A.; Khan, S.; Alofi, A. Damping Optimization and Energy Absorption of Mechanical Metamaterials for Enhanced Vibration Control Applications: A Critical Review. *Polymers* **2025**, *17*, 237. [CrossRef] [PubMed]
6. Plamadiala, I.; Croitoru, C.; Pop, M.A.; Roata, I.C. Enhancing Polylactic Acid (PLA) Performance: A Review of Additives in Fused Deposition Modelling (FDM) Filaments. *Polymers* **2025**, *17*, 191. [CrossRef] [PubMed]
7. Pinho, A.C.; Morais, P.V.; Pereira, M.F.; Piedade, A.P. Changes in the Antibacterial Performance of Polymer-Based Nanocomposites Induced by Additive Manufacturing Processing. *Polymers* **2025**, *17*, 171. [CrossRef] [PubMed]
8. Seifert, L.; Leuchtenberger-Engel, L.; Hopmann, C. Development of an Analytical Model for Predicting the Shear Viscosity of Polypropylene Compounds. *Polymers* **2025**, *17*, 126. [CrossRef] [PubMed]
9. Barera, G.; Pegoretti, A. Large-format 3D-printed carbon fiber-reinforced composites in tooling applications. *Prog. Addit. Manuf.* **2025**, *1*, 20. [CrossRef]

10. Lendvai, L.; Jakab, K.S.; Fekete, I.; Rigotti, D.; Pegoretti, A. Effect of filament humidity on the properties of material extrusion 3D-printed acrylonitrile butadiene styrene/hexagonal boron nitride composites. *Emergent Mater.* **2025**, *1*, 16. [CrossRef]
11. Coser, M.; Perin, D.; Fredi, G.; Aliotta, L.; Gigante, V.; Lazzeri, A.; Dorigato, A.; Pegoretti, A. Self-healing of polyamide 6/cyclic olefin copolymer/carbon fiber composites under quasi-static, impact, and fatigue conditions. *Compos. Sci. Technol.* **2025**, *268*, 111213. [CrossRef]
12. Enzo, R.P.D.; Bossio, G.R.; Bruno, M.M. Effect of FDM printing patterns on mechanical properties of ABS. *Rapid Prototyp. J.* **2023**, *29*, 2098–2108. [CrossRef]
13. Rahimipour, S.; Bahri-Laleh, N.; Ehsani, M.; Hedayati-Moghaddam, A.; Mokhtari-Aliabad, J.; Tabatabaei, S.S.; Mirmohammadi, S.A. Preparation and Properties of Enhanced Bio-Based PLA/PA6/Graphene Nanocomposites in the Presence of an Ester–Amide Exchange Catalyst. *J. Polym. Environ.* **2021**, *29*, 2302–2309. [CrossRef]
14. Sharma, S.S.; Singh, I.; Jani, S. Analysis of Temperature Effect for Fused Deposition Modeling-Printed Parts Using Acrylonitrile Butadiene Styrene Material in Additive Manufacturing. *J. Mater. Eng. Perform.* **2025**, *1*, 7. [CrossRef]
15. Rahul, P.; Jani, S.; Ankita, J. Review on multi-objective optimization of FDM process parameters for composite materials. *Int. J. Interact. Des. Manuf.* **2022**, *17*, 2115–2125. [CrossRef]
16. Dhand, A.P.; Davidson, M.D.; Zlotnick, H.M.; Kolibaba, T.J.; Killgore, J.P.; Burdick, J.A. AM of highly entangled polymer networks. *Science* **2024**, *1024*, 566–572. [CrossRef] [PubMed]
17. Gu, D.; Shi, X.; Poprawe, R.; Bourell, D.L.; Setchi, R.; Zhu, J. Material-structure-performance integrated laser-metal AM. *Science* **2021**, *1214*, 1487. [CrossRef] [PubMed]
18. Berman, A.; Hsiao, K.; Root, S.E.; Choi, H.; Ilyn, D.; Xu, C.; Stein, E.; Cutkosky, M.; DeSimone, J.M.; Bao, Z. Additively manufactured micro-lattice dielectrics for multiaxial capacitive sensors. *Sci. Adv.* **2024**, *2*, 8866. [CrossRef] [PubMed]
19. Mayer, F.; Richter, S.; Westhauser, J.; Blasco, E.; Barner-Kowollik, C.; Wegener, M. Multimaterial 3D laser microprinting using an integrated microfluidic system. *Sci. Adv.* **2019**, *2*, 9160. [CrossRef] [PubMed]
20. Yao, Y.; Shapiro, M.G. Using ultrasound to 3D-print materials. *Science* **2023**, *2*, 1126. [CrossRef] [PubMed]
21. Xie, R.; Shi, Y.; Liu, H.; Chen, S. A novel friction and rolling based solid-state AM method: Microstructure and mechanical properties evaluation. *Mater. Today Commun.* **2021**, *4*, 103005. [CrossRef]
22. Wu, W.; Xue, J.; Wang, L.; Zhang, Z.H.; Hu, Y.; Dong, C.W. Forming process, microstructure, and mechanical properties of thin-walled 316L stainless steel using speed-cold-welding AM. *Metals* **2019**, *3*, 109. [CrossRef]
23. Zhang, X.Y.; Fang, G.; Leeflang, S.; Zhou, J. Effect of subtransus heat treatment on the microstructure and mechanical properties of additively manufactured Ti-6Al-4V alloy. *J. Alloys Compd.* **2018**, *4*, 1562–1575. [CrossRef]
24. Yang, Y.Q.; Jiang, R.W.; Han, C.J.; Chen, J.Q.; Li, H.R.; Wang, Y.; Tang, J.R.; Zhou, H.; Hu, W.N.; Zheng, B.Y.; et al. Frontiers in Laser Additive Manufacturing Technology. *Addit. Manuf. Front.* **2024**, *746*, 200160. [CrossRef]
25. Ogi, K.; Nishikawa, T.; Okano, Y.; Taketa, I. Mechanical properties of ABS resin reinforced with recycled CFRP. *Adv. Compos. Mater.* **2007**, *4*, 181–194. [CrossRef]
26. Parpala, R.C.; Popescu, D.; Pupaza, C. Infill parameters influence over the natural frequencies of ABS specimens obtained by extrusion-based 3D printing. *Rapid Prototyp. J.* **2021**, *7*, 1273–1285. [CrossRef]
27. Qi, X.; Shen, X. Multidisciplinary design optimization of turbine disks based on ANSYS workbench platforms. *Procedia Eng.* **2015**, *1279*, 1275–1283. [CrossRef]
28. Saritha, G.; Iswarya, T.; Keerthana, D.; Baig, D.T. Micro universal testing machine system for material property measurement. *Mater. Today Proc.* **2023**, *21*, 329. [CrossRef]
29. Peleg, M. Texture profile analysis parameters obtained by an Instron universal testing machine. *J. Food Sci.* **1976**, *2*, 721–722. [CrossRef]
30. *GB/T 228*; Metallic Materials—Tensile Testing—Part 1: Method of Test at Room Temperature. Standardization Administration of the People's Republic of China: Beijing, China, 2021.
31. *ISO 6892-1*; Metallic Materials—Tensile Testing—Part 1: Method of Test at Room Temperature. International Organization for Standardization: Geneva, Switzerland, 2016.
32. Miled, K.; Roy, R.L.; Sab, K.; Boulay, C. Compressive behavior of an idealized EPS lightweight concrete: Size effects and failure mode. *Mech. Mater.* **2004**, *5*, 1031–1046. [CrossRef]
33. *ISO 16269-6*; Statistical Interpretation of Data—Part 6: Determination of Statistical Tolerance Intervals. International Organization for Standardization: Geneva, Switzerland, 2005.

polymers

MDPI

Article

Additive Manufacturing to Mimic the Nonlinear Mechanical Behavior of Cardiac Soft Tissue

Sara Valvez [1], M. Oliveira-Santos [2], L. Gonçalves [2], A. P. Piedade [1] and A. M. Amaro [1,*]

[1] University of Coimbra, Centre for Mechanical Engineering, Materials and Processes (CEMMPRE, ARISE), Department of Mechanical Engineering, 3030-788 Coimbra, Portugal; sara.valvez@dem.uc.pt (S.V.); ana.piedade@dem.uc.pt (A.P.P.)
[2] University of Coimbra, Institute for Clinical and Biomedical Research (iCBR), Faculty of Medicine, 3000-548 Coimbra, Portugal; manuel_ol_santos@hotmail.com (M.O.-S.); lgoncalv@ci.uc.pt (L.G.)
* Correspondence: ana.amaro@dem.uc.pt

Abstract

Soft biological tissues display highly nonlinear and anisotropic mechanical behavior, which is critical to their physiological function. Replicating these mechanical properties using engineered materials and additive manufacturing represents a significant challenge in biomedical engineering, particularly for surgical simulation, device development, and preclinical testing. The left atrial appendage (LAA) was selected since it plays a central role in thrombus formation during atrial fibrillation, significantly contributing to cardioembolic stroke. This study proposes a framework for reproducing the nonlinear stress–strain behavior of soft tissue using 3D-printed models. The methodology integrates experimental material selection with optimization of key printing parameters to ensure structural reliability and functional mechanical performance. Two polymers—polyurethane (TPU) and a thermoplastic with elastomer-type behavior (TPE)—were selected for their tunable hardness and elasticity. A parametric study was conducted to investigate the effects of Shore A hardness (60A to 100A), infill density (0% to 100%), and external shell number (zero to two) on the tensile performance of printed models. Mechanical testing was performed to extract stress–strain curves and evaluate the mechanical response. The practical implications of this study are significant, demonstrating the potential of additive manufacturing for anatomical reproduction and replicating functional mechanical properties in soft tissue models.

Keywords: additive manufacturing; nonlinear material behavior; soft tissue mechanics; parametric optimization; left atrial appendage

1. Introduction

Soft biological tissues are fundamental to the structural and functional integrity of physiological systems, particularly within the cardiovascular domain, where they are subjected to dynamic loading and large deformations. These tissues exhibit complex mechanical behavior characterized by nonlinearity, anisotropy, and viscoelasticity, typically represented by a J-shaped stress–strain response [1]. Accurately reproducing such behavior in synthetic analogs remains a significant challenge in engineering concepts, especially in developing models intended for surgical simulation, device testing, and computational analysis [2–4]. A clinically relevant example of such tissue is the left atrial appendage (LAA), a compliant cardiac structure recognized as the primary site of thrombus formation in patients with atrial fibrillation (AF) and thus a major contributor to cardioembolic stroke [5]. Owing to its critical role in thromboembolic risk, the LAA has become a principal target for deploying occlusion devices, which aim to mechanically isolate the appendage from systemic circulation and thereby reduce stroke incidence. Figure 1 illustrates the role of a synthetic LAA model in supporting occlusion planning, where the replication of patient-specific anatomy enables accurate device sizing and fit assessment.

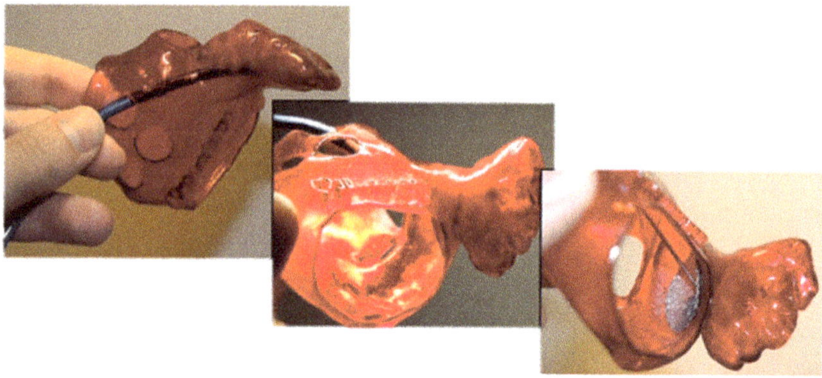

Figure 1. LAA occlusion planning: appropriate device sizing is supported by a synthetic model.

Although LAA models have already been integrated into interventional cardiac simulators, the literature reveals two predominant approaches. The first focuses primarily on anatomical reliability, aiming to reproduce the geometry of the appendage for visual and spatial assessment of device fit [6]. These models are typically made from rigid materials and neglect the mechanical behavior of native tissue. The second approach explores the use of flexible materials to mimic the tissue's mechanical properties [7]. However, inadequate replication of the tissue's stress–strain behavior may lead to unrealistic device–tissue interactions during simulation, resulting in incorrect estimation of device type, size, or positioning. If the mechanical properties and stress–strain behavior of the model do not accurately replicate those of the native tissue, device–tissue interaction during simulation may be unrealistic, potentially leading to incorrect assessment of device fit and anchoring. Consequently, it may compromise the effectiveness of training or planning and misrepresent potential clinical risks such as peri-device leakage, device embolization, incomplete closure, or excessive stress on the atrial wall [8]. These limitations underscore the need for a comprehensive mechanical characterization, particularly in terms of reproducing the nonlinear stress–strain response essential for simulating realistic device–tissue interactions.

Cardiac tissues, including the LAA, exhibit complex mechanical behavior characterized by anisotropy, nonlinear elasticity, and viscoelasticity [8]. Specifically, the LAA displays high strain capacity and distensibility, which are essential for its role as a decompression chamber and for modulating atrial function [9]. Strain capacity reflects tissue contractility and reduced strain is associated with impaired function and increased thrombus risk in atrial fibrillation [10]. Distensibility, the ability to accommodate volume under pressure, is linked to tissue elasticity and prevents blood stasis under hemodynamic load [11]. These functional demands translate into engineering requirements for materials capable of large, reversible deformations with minimal plasticity. In engineering terms, this behavior is characterized by a material's ability to undergo large reversible deformations within the elastic domain. Therefore, materials selected to mimic the LAA should exhibit high elasticity and preserve mechanical integrity at large strains. These requirements exclude materials with limited deformability, or predominant plastic behavior. Accordingly, elastomeric materials are the most suitable candidates for mimicking the mechanics of LAA tissue. Their inherent flexibility, ability to recover after deformation, and tunable Shore hardness make them ideal for reproducing the nonlinear, highly elastic response of native cardiac tissue [12]. The literature on this subject mentions the use of polymeric materials and additive manufacturing (AM) technologies to create customized, patient-specific models of the LAA [7,13]. Among the published papers, the use of flexible materials to produce these models was associated with better procedural outcomes [13]. Materials with rubber-like properties, such as Tango® (from Stratasys Ltd., Eden Prairie, MN, USA) [6,14–16], polyurethane (TPU) in filament form, gelatin [17], silicone [18,19] and Latex [20] in the liquid state, are reported in the literature. Although flexible materials have shown improved realism in procedural outcomes, a significant number of these studies lack specification of the materials used or fail to provide systematic mechanical characterization [21–24]. In other works, one from the authors, it was found that thermoplastic polyurethane (TPU) exhibited mechanical properties closer to those of LAA tissue than other tested materials [12, 25]. However, deviations in the stress–strain response, particularly in curve shape and modulus, highlighted the need for further refinement.

Since quantitative data on the mechanical behavior of biological LAA tissue are not widely available in the published literature, this study adopts the only experimentally obtained stress–strain curve identified, to the best of the authors' knowledge, as a reference benchmark for model evaluation (Figure 2) [1].

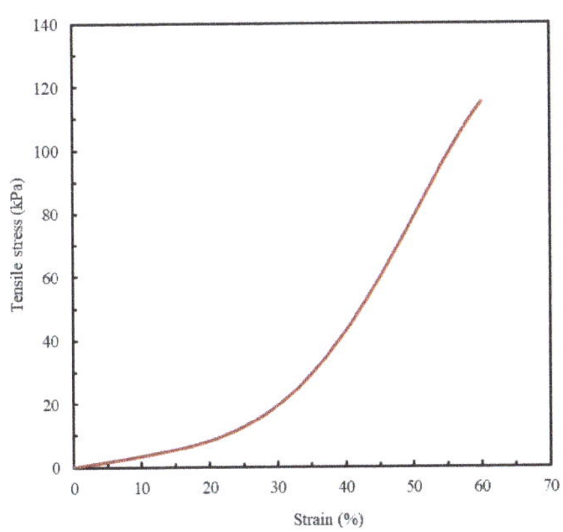

Figure 2. LAA occlusion planning: appropriate device sizing is supported by a synthetic model. Tensile stress- strain curve. Adapted from [1].

In addition to material selection, the manufacturing process plays a decisive role in determining final mechanical and surface characteristics. Given the LAA's anatomically complex and highly variable morphology, conventional manufacturing methods fall short in replicating its geometry and mechanical functionality. In this context, AM offers significant advantages, including reproducing complex anatomical geometries and fine-tuning mechanical properties through precise control of processing parameters. While stereolithography (SLA) offers smooth, high-resolution parts, fused filament fabrication (FFF) allows greater material flexibility and produces textured surfaces more aligned with the morphology of cardiac tissues [26]. For these reasons, FFF was selected in this study to fabricate tissue-mimicking specimens.

Although AM has advanced in biomedical applications, systematic efforts to optimize material-process combinations for replicating the nonlinear tensile behavior of cardiac tissues, particularly the LAA, remain limited, especially regarding mechanical properties.

This study aims to fill that gap by developing a framework capable of replicating the nonlinear mechanical behavior of LAA biological tissue. A parametric approach is employed to investigate the influence of key parameters that significantly affect structural integrity, namely Shore A hardness, infill density, and shell number, on the tensile response of FFF-printed elastomeric specimens. The goal is to identify optimized material–structure combinations suitable for future implementation in anatomically accurate LAA models, enhancing the realism of mechanical behaviors and the functional relevance of cardiovascular simulators and supporting more effective, patient-specific device evaluation.

2. Materials and Methods

2.1. Materials

Thermoplastic polymers with elastomeric-like behavior (TPU and TPE) were selected based on their Shore hardness, specifically TPU 60D, TPU 93A, TPU 40D, TPU 85A, TPE 70A, and TPE 60A. To enable a direct comparative analysis, hardness values in Shore D values were converted to Shore A, as this scale is more appropriate for measuring the hardness values of softer materials. Therefore, throughout this study, TPU 60D and TPU 40D will henceforth be referred to as TPU 100A and TPU 90A, respectively. The polymers were purchased from Filament2Print (Pontevedra, Spain) in filament form, with a diameter of 1.75 ± 0.04 mm.

2.2. Material Selection and Optimization of Printing Parameters

Since the primary objective of this study is to replicate the mechanical behavior of biological LAA tissue, it is essential to reproduce its anisotropic nature, a fundamental characteristic of cardiac tissue [27]. Cardiac muscle fibers are arranged in helical and oblique layers, which result in multidirectional load paths not limited to the Cartesian 0° and 90° axes [28]. Consequently, to mimic this structure, a 45°/−45° infill orientation was applied, allowing a more realistic simulation of fiber alignment and mechanical behavior, as illustrated in Figure 3.

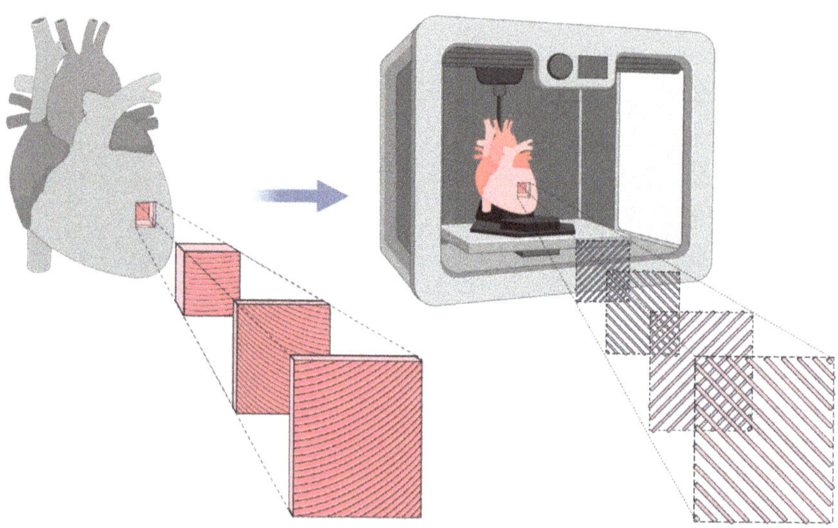

Figure 3. Representation of cardiac tissue anisotropy and its biomimetic counterpart fabricated using FFF. The biomimetic model utilizes a 45°/−45° infill pattern to replicate the structural alignment of natural cardiac tissue. Adapted from [29].

Although the 45°/−45° pattern does not replicate the full helicoidal complexity of the cardiac tissue, it offers mechanical advantages over 0°/90° configurations by improving stress distribution, reducing localized failure, and enhancing fatigue resistance. This infill strategy provides a more uniform mechanical response, better mimicking the anisotropic nature of the LAA under physiological loads [30].

Healthy cardiac tissue is typically orthotropic, with stiffness varying in three orthogonal directions based on the alignment of fibers, laminar sheets, and the extracellular matrix [31]. In contrast, the LAA in patients with atrial fibrillation often exhibits anisotropic behavior due to fibrotic remodeling, resulting in directional mechanical heterogeneity [32]. Therefore, the aim is not to reproduce an ideal orthotropic model, but rather a controlled anisotropic response aligned with pathological LAA tissue mechanics.

The 45°/−45° filament orientation introduces directional reinforcement, simulating the tensile, compressive, and torsional stresses encountered in vivo. This pattern reduces stress concentration and improves resistance to multi-axial loads, which is especially relevant for models subjected to cyclic loading conditions. Compared to 0°/90°, it offers a mechanical compromise across multiple directions, approximating the elastic and fatigue behavior of native LAA tissue [33,34]. Thus, the selected infill strategy supports the development of more mechanically faithful models for use in simulation-based planning of LAA occlusion procedures.

In the first stage of this study, the material selection process, the specimens were fabricated using an original Prusa MK4S (Prusa Research, Prague, Czech Republic) and fixed printing parameters, including a layer height of 0.2 mm, two shells, 100% infill density, and an infill pattern of lines oriented at +45°/−45°. The remaining parameters varied according to the values presented in Table 1 and the supplier's recommendations.

Table 1. Principal printing parameters for each material.

Material	Extrusion Temperature (°C)	Bed Temperature (°C)	Printing Speed ($mm \cdot s^{-1}$)
TPU 100A	210	40	40
TPU 93A	225	50	40
TPU 90A	220	60	40
TPU 85A	230	60	40
TPE 70A	230	60	40
TPE 60A	220	30	60

The second stage of the research aimed to evaluate the impact of the two most critical printing parameters on the structural integrity of AM models: the number of shells and the infill density [35,36]. Figure 4 schematizes the combination of the printing parameters used in addition to the fixed parameters of the first part.

Variations in the printing parameters were intended to assess the impacts of the shell number (zero, one, and two) and infill density (D) (0%, 25%, 50%, and 100%) on the mechanical properties of specimens. Shell number is defined as equal to the perimeter contours (Np), and in this study, the numbers of top solid layers (T) and bottom solid layers (B) were set to be equal to the shell number (i.e., shell number = Np = T = B). D was specified independently. Specimens were fabricated with 0, 1, or 2 shells; accordingly: 0 shell (infill-only: Np = 0, T = 0, B = 0, D > 0), 1 shell (Np = 1, T = 1, B = 1), and 2 shells (Np = 2, T = 2, B = 2). Multiple samples were fabricated for each selected material and set of printing parameters to ensure at least five valid test results for each material–printing-parameter combination.

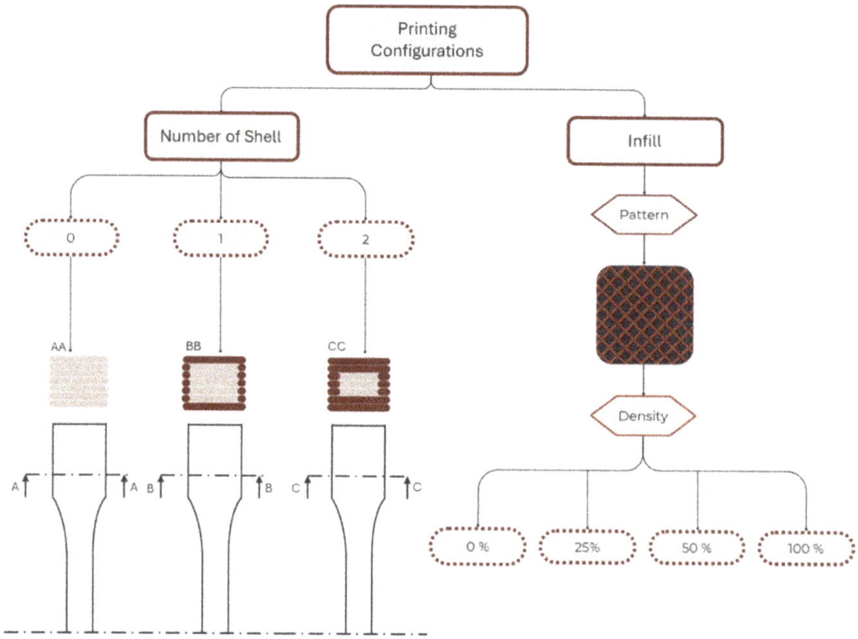

Figure 4. Schematic representation of the printing parameters used in the second part of the study.

2.3. Mechanical Characterization

When studying mechanical properties for designing cardiovascular medical simulators, it is essential to emphasize the role of radial resistance and the material's ability to anchor an occlusion device. Radial resistance is described as the capacity of a material to withstand deformation when subjected to compressive loads from multiple directions simultaneously [37]. It is considered a critical characteristic for accurately mimicking the function of the LAA, since this biological structure undergoes significant deformations and sustains pressures during the cardiac cycle [9]. Therefore, radial resistance emerges as a critical mechanical property in the planning and production of LAA models. The existing literature suggests that tensile tests provide a means to evaluate radial strength and stiffness, with established direct correlations between these parameters and other material properties, such as ultimate stress (σ) and Young's modulus (E), respectively [38]. Additionally, since the results used as a control for the current study were derived from a tensile stress–strain curve (Figure 2) [1], tensile tests were selected as the most appropriate characterization technique.

The specimens were produced and characterized using tensile tests in accordance with the ISO 527 standard [39]. Figure 5a illustrates the geometry used for the samples in the experiments. For each material, at least five specimens were produced, following the configuration shown in Figure 5a, to ensure repeatability and reproducibility in the analyses performed. Figure 5b depicts the experimental setup, which utilized Shimadzu Autograph AGS-X universal tensile test equipment equipped with a 5 kN load cell and associated with TrapeziumX software (Version 1.5.1). The tests were conducted at a grip speed of 5 mm/min under room temperature conditions.

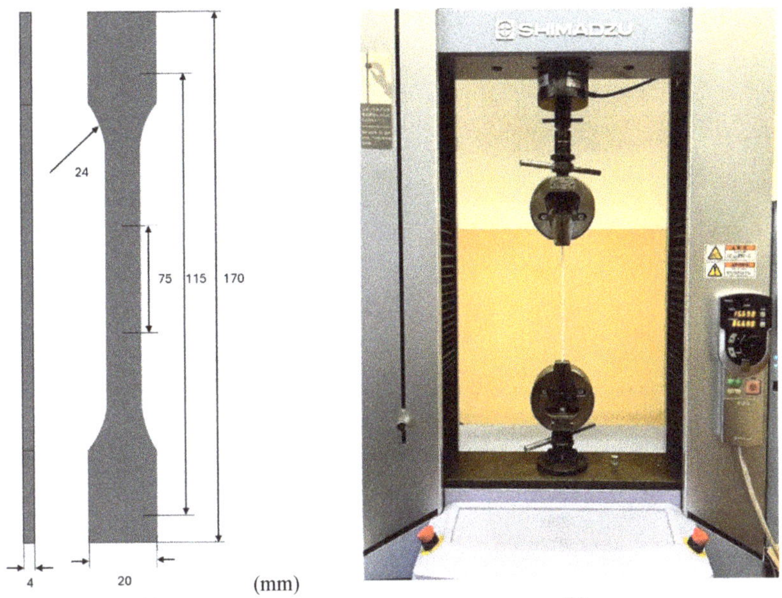

(**a**) (**b**)

Figure 5. (**a**) Schematic representation of the samples according to ISO 527 standard [39]. (**b**) Experimental tensile test setup using Shimadzu Autograph AGS-X.

The stress–strain curves obtained from the tensile tests were analyzed to determine the critical mechanical properties for replicating LAA tissue's mechanical behavior, including the maximum tensile stress (σ), Young's modulus (E), and maximum strain (ε). The study will compare the stress–strain behavior of the printed specimens to the documented mechanical behavior of LAA biological tissue reported in the literature to validate the suitability of the results (Figure 2).

3. Results and Discussion

This section, divided into three subsections, presents and discusses the results regarding material selection, the influence of the number of shells and infill density, and the replication of the mechanical behavior of the LAA biological tissue.

3.1. Material Selection

The first experimental tests were conducted to determine the tensile properties of the six polymers referred to before: TPU 100A, TPU 93A, TPU 90A, TPU 85A, TPE 70A, and TPE 60A. At least five valid tests were conducted for each combination of material and printing parameters, with the resulting strain-stress curves demonstrating a high level of repeatability. Figure 6 presents the average stress–strain curves for each material, with each sample produced using two shells and 100% infill density.

Table 2 displays the average and standard deviation values of the tensile properties of the six pre-selected materials, particularly the maximum tensile stress (σ) and Young's modulus (E).

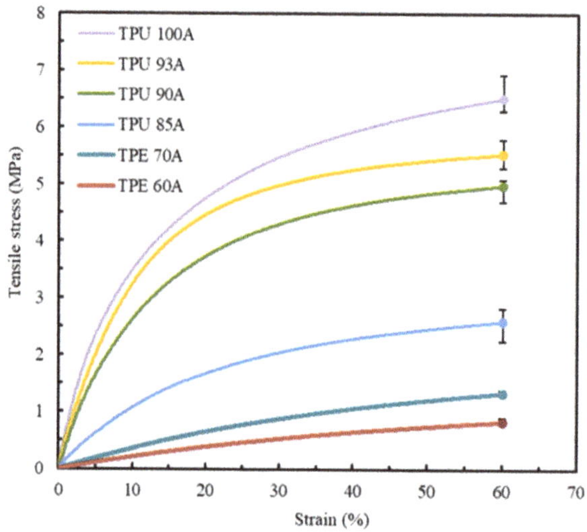

Figure 6. Average stress–strain tensile curves of all the synthetic polymers studied, with two shells and 100% infill.

Table 2. Average and standard deviation values of σ and E of the printed materials using two shells and 100% infill.

Material	σ (MPa)	E (kPa)
TPU 100A	6.8 ± 0.7	409.6 ± 19.6
TPU 93A	5.6 ± 0.6	369.8 ± 59.7
TPU 90A	5.0 ± 0.6	309.9 ± 60.2
TPU 85A	2.6 ± 0.8	113.9 ± 5.8
TPE 70A	1.3 ± 0.2	36.3 ± 1.9
TPE 60A	0.8 ± 0.1	20.6 ± 1.9

A strain limit of 60% was established as the comparative benchmark for all analyzed materials. This value is aligned with a study that reported that the stress–strain curve of the LAA terminates at a similar strain value [1].

The low variability of the results is consistent across all the materials analyzed, as shown by the low dispersion represented in Figure 6 and Table 2. The stress–strain curves of the materials exhibit different mechanical behaviors, with all materials demonstrating higher stress at a strain of 60%. TPU materials show an ascending convex curve, indicating a significant increase in tensile stress with increasing strain. Additionally, it is possible to observe that all the TPU materials reveal a higher strain resistance when compared to TPE. TPU 100A, TPU 93A, and TPU 90A exhibit a steeper stress–strain curve with a maximum stress of around 6.8 MPa, 5.6 MPa, and 5 MPa, respectively. On the other hand, TPU 85A, TPE 70A, and TPE 60A exhibit flatter curves, which show lower tensile stress and stiffness and, consequently, higher deformation capacities. While TPU 85A can achieve a maximum tensile stress value of approximately 2.7 MPa, TPE 70A only reaches 1.3 MPa, a difference of around 52%. Among all the tested materials, TPE 60A shows the lowest stress of about 1 MPa and, according to its stress–strain curve evolution, the lowest Young's modulus. The results indicate that none of the combinations of material-processing parameters allow for an accurate and direct replication of the mechanical behavior of LAA biological tissue. Consequently, aiming to attain the tensile values and the geometry of the stress–strain curve of the control specimen, only TPU 85A, TPE 70A, and TPE 60A were selected for further study.

3.2. Influence of the Number of Shells and Infill Density

Focusing on the pre-selected materials TPU 85A, TPE 70A, and TPE 60A according to the results presented in Section 3.1, only two printing parameters were varied, the number of shells and infill density, with the aim of mimicking the mechanical behavior of biological tissues [12,40]. Figure 7a–c display the tensile stress–strain curves of TPU 85A, TPE 70A, and TPE 60A, respectively. "1S" and "2S" refer to the number of shells (one and two, respectively), while the percentages (0%, 25%, 50%, and 100%) indicate the infill density.

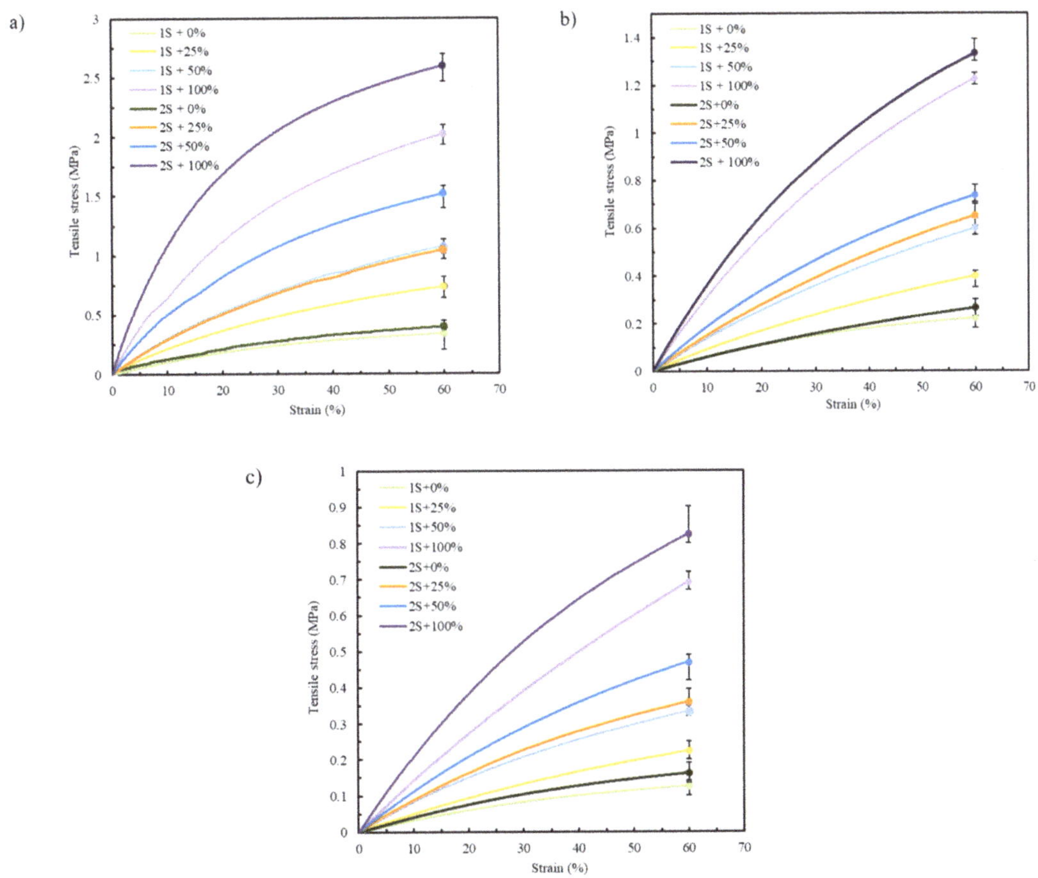

Figure 7. Average stress–strain tensile curves for (**a**) TPU 85A, (**b**) TPE 70A, and (**c**) TPE 60A.

Table 3 displays the stress (σ) and Young's modulus (E) of TPU 85A, TPE 70A, and TPE 60A considering the previously mentioned variations in printing parameters, specifically in terms of shells and infill density.

Table 3. Average and standard deviation values of σ and E of TPU 85A, TPE 70A, and TPE 60A, according to different sets of printing parameters.

Material	Printing Parameters	σ (MPa)	E (kPa)
TPU 85A	2S + 100%	2590 ± 800	113.9 ± 5.8
	2S + 50%	1520 ± 74	53.5 ± 8.7
	2S + 25%	1040 ± 14	28.4 ± 1.2
	2S + 0%	390 ± 40	10.9 ± 1.2
	1S + 100%	2022 ± 300	73.6 ± 5.2
	1S + 50%	1080 ± 124	29.4 ± 3.4
	1S + 25%	730 ± 90	23.1 ± 1.3
	1S + 0%	330 ± 60	9.6 ± 0.7
TPE 70A	2S + 100%	1331 ± 154	36.3 ± 1.9
	2S + 50%	736 ± 95	18.7 ± 2.2
	2S + 25%	650 ± 49	15.3 ± 1.5
	2S + 0%	265 ± 5	6.2 ± 0.5
	1S + 100%	1226 ± 48	33.5 ± 1.1
	1S + 50%	597 ± 4	14.5 ± 1
	1S + 25%	397 ± 7	9.1 ± 0.1
	1S + 0%	220 ± 3	6.1 ± 0.7
TPE 60A	2S + 100%	825 ± 30	20.6 ± 1.9
	2S + 50%	412 ± 30	11.8 ± 0.9
	2S + 25%	362 ± 18	8.7 ± 0.8
	2S + 0%	222 ± 40	4.0 ± 0.3
	1S + 100%	469 ± 30	14.1 ± 2.3
	1S + 50%	333 ± 20	8.1 ± 0.7
	1S + 25%	253 ± 0.021	4.9 ± 0.2
	1S + 0%	125 ± 0.05	3.1 ± 0.1

According to the results in Figure 7 and Table 3, the mechanical behavior of each printed sample was strongly influenced by the selected printing parameters, regardless of the material. This highlights the critical role of these parameters in determining the structural integrity of the printed specimens. The literature consistently reports that shell number and infill density are among the most influential parameters affecting the mechanical performance of FFF-printed parts [41]. Specifically, increasing the number of outer shells leads to higher tensile strength, as the shell region functions as the primary load-bearing structure [42,43].

3.2.1. Shell Number

A greater shell number results in denser and mechanically more stable specimens, improving their capacity to withstand tensile loading [36]. However, there appears to be a saturation point beyond which further increases in shell number do not translate into significant gains in strength, potentially due to manufacturing-related defects or internal stress concentrations introduced during the printing process [44]. Moreover, shell continuity is essential: interrupted or poorly bonded shells can significantly reduce a specimen's ability to carry loads, even when other parameters are favorable [45].

Within the range of shells tested (one and two), the findings of the current study corroborate previous studies [35,44,46], confirming that a greater shell number resulted in a clear and consistent impact on the tensile behavior of the specimens. This reinforces its role as a critical parameter in optimizing the mechanical performance of structures produced via FFF [35,36].

In Figure 7, specimens printed with 2S consistently exhibit higher tensile strength compared to those with 1S. This tendency can be attributed to the increased structural stiffness provided by the additional shell material. Given that the shells are oriented longitudinally to the direction of tensile loading, they offer enhanced resistance, allowing for more effective load transfer along the length of the specimen. This behavior reflects the intrinsic advantage of filament alignment in the load direction, which promotes higher strength and stiffness under uniaxial stress [47].

In the case of TPU 85A, the difference between 1S and 2S becomes more pronounced as strain values increase, with 2S exhibiting a steeper curve, reaching tensile stress values higher than 2.6 MPa in samples with 100% infill density. A similar tendency is observed in the specimens printed using TPE 70A and TPE 60A, although they have lower tensile stress values due to their more ductile behavior. TPE 70A and TPE 60A samples with 2S and 100% infill show tensile stress values of approximately 1.33 MPa and 0.75 MPa, respectively. These values contrast with those with only 1S, which exhibit tensile stresses around 1.26 MPa and 0.5 MPa, respectively, highlighting their reduced ability to withstand higher stress values, which means a decrease of about 5.3% and 33.3% when comparing the 2S and 1S specimens for TPE 70A and TPE 60A, respectively.

3.2.2. Infill Density

The second parameter evaluated, infill density, also significantly affects the mechanical response of FFF-printed parts [48]. Higher infill densities generally increase tensile strength and stiffness, while lower densities promote more flexible behavior, which the results support. TPU 85A specimens showed the highest tensile strength at greater infill percentages in this study, due to improved internal load distribution [49]. The same trend was observed in TPE 70A and TPE 60A, although their lower hardness resulted in lower stress values overall.

As shown in Figure 7, while all tested materials tolerated large deformations (up to 60% strain), neither the stress–strain curves nor the maximum stress values fully replicated the nonlinear mechanical behavior of native LAA tissue. Accordingly, further optimization is required to achieve accurate mechanical biomimicry.

3.3. LAA Tissue Mechanical Behavior Replication

In agreement with the literature and as confirmed by the results presented in Figure 7, the number of shells has a pronounced influence on structural integrity and stress–strain curve geometry [50,51]. However, the mechanical behavior of LAA biological tissue was not achieved with either of the selected combinations of printing parameters. Considering that both one and two shells offered greater structural integrity to the samples, the decision was made to exclude shells in subsequent analyses. Therefore, the next phase of this study will be conducted using zero shells to further investigate the material's mechanical properties without the influence of outer layers, focusing on examining the impact of infill density.

Building upon preliminary tests conducted with 0%, 25%, 50%, and 100% infill, a 25% density emerged as a promising compromise between mechanical performance and material efficiency, particularly when using zero shells. However, to further explore the lower limits of structural integrity while preserving printability, a 10% infill density was additionally introduced. As a result, two configurations were selected for analysis: zero shells with 10% infill (0S + 10%) and zero shells with 25% infill (0S + 25%). These tests aimed to evaluate the sensitivity of the material's mechanical response to internal structure variations under minimal reinforcement conditions.

Figure 8 shows a comparative analysis of the mechanical behavior of TPU 85A, TPE 70A, and TPE 60A with different infill density percentages (10% and 25%) and biological LAA tissue, which was considered the reference for this study [1].

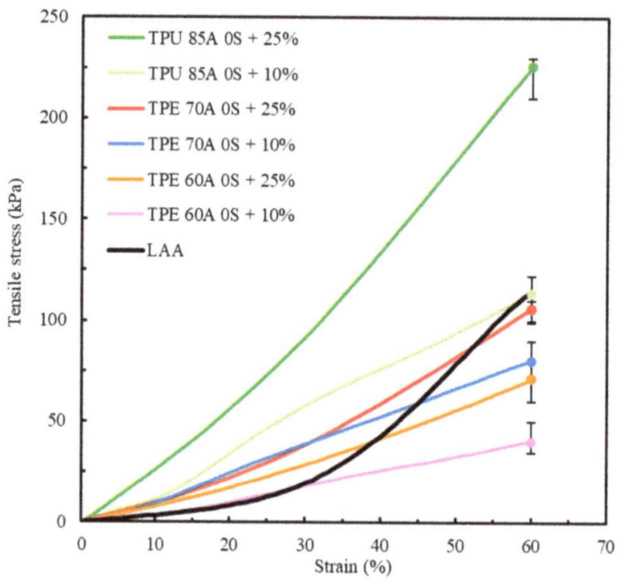

Figure 8. Comparison of the average stress–strain tensile curves between TPU 85A, TPE 70A and TPE 60A (this study) and LAA biological tissue [1].

Table 4 displays the maximum σ and E values of TPU 85A, TPE 70A, and TPE 60A for samples with the printing parameters 0S + 10% and 0% + 25%.

Table 4. Average and standard deviation values of σ and E of TPU 85A, TPE 70A, and TPE 60A for different printing parameter combinations.

Material	Printing Parameters	σ (kPa)	E (kPa)
TPU 85A	0S + 10%	114 ± 21	2.7 ± 0.3
	0S + 25%	226 ± 25	1.1 ± 0.2
TPE 70A	0S + 10%	80 ± 3	1.0 ± 0.1
	0S + 25%	106 ± 11	1.0 ± 0.1
TPE 60A	0S + 10%	40 ± 2	0.8 ± 0.1
	0S + 25%	72 ± 5	0.3 ± 0.1

The mechanical properties of cardiac biological tissues vary significantly depending on the tissue type (e.g., myocardium, cardiac valves, pericardium, etc.) and the experimental conditions (e.g., physiological state, stretch direction, etc.) [52–54]. Despite these varying conditions, the mechanical properties of these tissue types predominantly fall within the kPa range, maintaining consistency in their order of magnitude [55].

Matching the mechanical behavior of the synthetic LAA model to that of the biological tissue is essential and well documented in the literature [7]. Properties such as stiffness and compliance directly affect the interaction between the occlusion device and LAA wall, influencing procedural outcomes. The LAA exhibits a distinctive nonlinear stress–strain response, typically represented by a J-shaped curve [7,56]. This curve, characteristic of biological soft tissues, enables the tissue to remain compliant under low loads and progressively stiffen under increasing strain—an essential feature for energy absorption and structural integrity. This behavior is attributed to the native fiber-reinforced architecture of the extracellular matrix, which governs the tissue's mechanical adaptation [1,12]. Therefore, for a synthetic model to accurately replicate LAA mechanics, the selected material must reproduce not only the strain capacity and strength but also the overall geometry of the stress–strain curve.

The results shown in Figure 8 indicate a clear divergence between the mechanical behavior of TPU 85A and TPE 60A and LAA biological tissue. TPU 85A exhibited a significantly higher response to tensile stress, particularly in samples with 25% infill, reaching approximately 226 kPa at 60% strain. This represents a significantly higher stress value, approximately 46.9% greater than that of LAA tissue, which does not exceed 120 kPa. TPU 85A's mechanical behavior suggests high stiffness, which makes it inappropriate for applications that intend to replicate the flexibility of biological tissues such as LAA tissues.

At a 10% infill, the TPU 85A material still exhibited substantially higher tensile stress than LAA tissue. Conversely, TPE 60A exhibited a more flexible behavior at both infill density percentages, with a maximum tensile stress of 80 kPa (10% infill) and 106 kPa (25% infill) at 60% strain. However, despite presenting similar values in terms of strain, TPE still exhibited higher tensile stress than LAA biological tissue, especially in 25% infill density percentage samples. Even in 10% infill density samples, despite having the stress–strain curve which most resembles that of LAA tissue, it still displayed a considered difference in tensile stress, suggesting that the material is suitable for replicating the mechanical behavior of the LAA but not suitable when considering the values of its mechanical properties.

The results for TPE 70A for both infill variations, for the same number of shells, 0S + 10% and 0S + 25%, indicate that this material has an intermediary mechanical behavior compared to the other tested specimens. TPE 70A with 0S + 25% demonstrated a good replication of the profile of the stress–strain curve of LAA tissue, particularly until 40% deformation. In this strain range, TPE 70A with 0S + 25% exhibited higher tensile stress values than the LAA, reflecting greater stiffness relative to the target tissue. Despite this, the stress–strain curve of TPE 70A with 0S + 25% still shows a reasonable correlation with the characteristic stress–strain behavior of the LAA, particularly in terms of its overall shape and trend within this strain range. However, it is possible to observe that, from 40% strain onwards, there is an alteration in the mechanical behavior, indicating a different tendency towards higher strain values while exhibiting lower tensile stress.

On the other hand, regarding the overall strain range, TPE 70A samples with 0S + 10% presented lower tensile stress and stiffness values. Regardless of both TPE 70A infill variations exhibiting similarities in their stress–strain curves up to approximately 25% strain, beyond this point, samples with 0S + 10% exhibited an increase in stress–strain values, although with a reduced rate of progression. It is important to mention that, at 60% strain, both specimens of TPE 70A (0S + 10% and 0S + 25%) exhibited maximum stress values lower than those of LAA tissue, suggesting that the material is mechanically less resistant than the biological tissue.

Among all the parameters studied, the results highlight the adjustment capacity of the selected materials, given that variations in infill density resulted in distinct mechanical responses. Both TPU 85A and TPE 60A exhibit significantly steeper curves, indicating that these materials with the tested printing parameters are not suitable for replicating the behavior of LAA. Regarding TPE 70A, despite its divergence from the final part of the LAA's curve, particularly for the 0S + 25% printing parameters, its stress–strain behavior until 40% strain was similar, and its maximum tensile stress value was the closest, with a difference of 13.2%. The major differences that were observed concerning the nonlinear response at higher strain levels highlight the need for small changes in manufacturing conditions. These adjustments can improve the alignment with the behavior of biological tissue, especially at higher strain values.

The stress–strain curve considered for comparison (Figure 2) is the stress–strain curve of a healthy LAA. However, according to the literature, the LAA is more prone to blood clot generation in patients predisposed to atrial fibrillation episodes. The tissues of these LAAs typically exhibit stiffer and less linear mechanical behaviors [57]. Among the tested materials, TPE 70A's characteristics most closely resemble these characteristics. As our idea was to create three-dimensional models to allow physicians to plan and practice LAA occlusion procedures, selecting a higher infill density percentage is advantageous. This choice ensures that the occlusion device can be more effectively anchored, mimicking the mechanical properties required for clinical success. Therefore, the material that best replicates the mechanical behavior of LAA tissue is TPE 70A printed with zero shells, 25% infill, and a linear $45°/-45°$ infill pattern.

4. Conclusions

This research aimed to identify synthetic materials and printing parameters to mimic the mechanical behavior of LAA biological tissue. Among the selected materials, TPU and TPE, tensile test results indicated that the most suitable materials for this study's objective were the ones with lower hardness values, TPU 85A, TPE 70A, and TPE 60A, as they exhibited deformation values closer to those of the LAA. However, the results were not ideal, making it necessary to study several printing parameters.

Shell number (zero, one, and two) and infill density (0%, 10%, 25%, 50%, and 100%) were analyzed. It was concluded that, regardless of the printing parameters, TPU 85A, due to its higher stress–strain response, demonstrates characteristics that make it unsuitable for applications requiring greater flexibility, as typically observed in biological tissues. On the other hand, TPE 60A is a more flexible material and exhibits lower tensile values than LAA biological tissue. For samples with zero shells and 10% infill density, TPE 60A can accurately replicate the mechanical behavior of LAA up to 30% strain. Nonetheless, the stress–strain curves diverge for higher strain values, with TPE 60A showing lower tensile stress values than LAA tissue. Regarding TPE 70A, specimens printed with 0S + 10% exhibited some initial similarities to the mechanical behavior of LAA tissue, particularly at strain values lower than 26% when comparing the stress–strain curves of 0S + 10% and LAA biological tissue. However, in addition to exhibiting higher maximum stress values than those of the biological tissue throughout the entire analyzed range, its stress–strain curve reveals a mechanical behavior which was less representative of LAA tissue at higher strain values. Additionally, the samples produced with these printing parameters show a tendency for lower stiffness compared to the 0S + 25% ones, which can reduce their applicability for models that aim to simulate pathological conditions of the LAA (where the tissue behaves more stiffly and in a less linear manner). Furthermore, the anchoring process of occlusion devices can be compromised in models based on these printing parameters due to low structural density.

TPE 70A with the 0S + 25% configuration was identified as the most suitable option to replicate the mechanical characteristics of LAA tissue under pathological conditions, such as those observed in patients predisposed to clot formation during episodes of atrial fibrillation. Although the material displays higher tensile stress values than LAA tissue, its stress–strain behavior is the closest compared to the other evaluated materials, especially at strain values lower than 40%. Moreover, a 25% infill enables more effective anchoring of the occlusion device, allowing for more reliability in LAA occlusion procedural planning. Therefore, TPE 70A with 0S + 25% infill density was selected for developing LAA tridimensional models due to its capacity to attend to the mechanical behavior characteristics of LAA tissue while simultaneously facilitating the clinical practice of occlusion procedures.

This study marks a significant advancement in identifying materials suitable for mimicking the LAA. In fact, although some studies focus on the LAA, most are numerical, providing little to no insight into the mechanical properties of used materials.

5. Future Research and Limitations of the Study

The main results of this study are promising since the stress–strain curves of the synthetic materials closely resemble the mechanical response of LAA biological tissue. However, for accurate replication of this behavior, the study highlights the necessity of approaching novel infill patterns, maintaining selection of materials with zero shells and low infill density values. By isolating the infill pattern as a variable of interest, it is possible to better understand its impact on these materials' mechanical properties and overall effectiveness. This method could result in improved material designs that closely resemble the properties of LAA tissue and enhance progress in creating biomimetic structures for medical purposes.

As a limitation of this research, the lack of studies regarding the mechanical characterization of LAA biological tissues is underlined. The comparison between biological tissue and the synthetic materials was performed based only on a single LAA stress–strain curve. Despite the expected range of values of mechanical properties, LAA biological tissue characteristics are individual since they are affected by several factors, such as associated cardiac pathology and physical activity. So, the use of one LAA tensile stress–strain curve led to preliminary results.

Author Contributions: S.V.: methodology, validation, investigation, data curation, writing—original draft, writing—review and editing, visualization. M.O.-S.: conceptualization, writing—review and editing. L.G.: conceptualization, resources, supervision, writing—review and editing. A.P.P.: conceptualization, methodology, validation, writing—review and editing, visualization, supervision. A.M.A.: conceptualization, methodology, validation, resources, writing—review and editing, visualization, supervision. All authors have read and agreed to the published version of the manuscript.

Funding: Sara Valvez acknowledges Fundação para a Ciência e Tecnologia (FCT), Portugal, for the financial support through the PhD Grant 2023.00413.BD.

Institutional Review Board Statement: Not applicable.

Data Availability Statement: The original contributions presented in this study are included in the article. Further inquiries can be directed to the corresponding authors.

Acknowledgments: This research is sponsored by national funds through FCT—Fundação para a Ciência e a Tecnologia, under projects UID/00285—Centre for Mechanical Engineering, Materials and Processes and LA/P/0112/2020. During the preparation of this manuscript/study, the author(s) used

Grammarly, online subscription version for the purposes of improving the readability and language of the manuscript. The authors have reviewed and edited the output and take full responsibility for the content of this publication.

Conflicts of Interest: The authors declare no conflicts of interest.

References

1. Lin, C.; Liu, L.; Liu, Y.; Leng, J. 4D Printing of Bioinspired Absorbable Left Atrial Appendage Occluders: A Proof-of-Concept Study. *ACS Appl. Mater. Interfaces* **2021**, *13*, 12668–12678. [CrossRef]
2. Gamberini, G.; Maglio, S.; Mariani, A.; Mazzotta, A.D.; Forgione, A.; Marescaux, J.; Melfi, F.; Tognarelli, S.; Menciassi, A. Design and Preliminary Validation of a High-Fidelity Vascular Simulator for Robot-Assisted Manipulation. *Sci. Rep.* **2024**, *14*, 4779. [CrossRef] [PubMed]
3. DeCampos, D.; Teixeira, R.; Saleiro, C.; Oliveira-Santos, M.; Paiva, L.; Costa, M.; Botelho, A.; Gonçalves, L. 3D Printing for Left Atrial Appendage Closure: A Meta-Analysis and Systematic Review. *Int. J. Cardiol.* **2022**, *356*, 38–43. [CrossRef]
4. Zhalmuratova, D.; La, T.-G.; Yu, K.T.-T.; Szojka, A.R.A.; Andrews, S.H.J.; Adesida, A.B.; Kim, C.; Nobes, D.S.; Freed, D.H.; Chung, H.-J. Mimicking "J-Shaped" and Anisotropic Stress–Strain Behavior of Human and Porcine Aorta by Fabric-Reinforced Elastomer Composites. *ACS Appl. Mater. Interfaces* **2019**, *11*, 33323–33335. [CrossRef]
5. Sulague, R.M.; Whitham, T.; Danganan, L.M.L.; Effiom, V.; Candelario, K.; Latif, N.; Hameed, I. The Left Atrial Appendage and Atrial Fibrillation—A Contemporary Review. *J. Clin. Med.* **2023**, *12*, 6909. [CrossRef]
6. Goitein, O.; Fink, N.; Guetta, V.; Beinart, R.; Brodov, Y.; Konen, E.; Goitein, D.; Di Segni, E.; Grupper, A.; Glikson, M. Printed MDCT 3D Models for Prediction of Left Atrial Appendage (LAA) Occluder Device Size: A Feasibility Study. *EuroIntervention* **2017**, *13*, e1076–e1079. [CrossRef] [PubMed]
7. Valvez, S.; Oliveira-Santos, M.; Gonçalves, L.; Amaro, A.M.; Piedade, A.P. Preprocedural Planning of Left Atrial Appendage Occlusion: A Review of the Use of Additive Manufacturing. *3D Print. Addit. Manuf.* **2024**, *11*, 333–346. [CrossRef]
8. Misra, S.; Ramesh, K.T.; Okamura, A.M. Modelling of Non-Linear Elastic Tissues for Surgical Simulation. *Comput. Methods Biomech. Biomed. Engin* **2010**, *13*, 811–818. [CrossRef]
9. Al-Saady, N.M.; Obel, O.A.; Camm, A.J. Left Atrial Appendage: Structure, Function, and Role in Thromboembolism. *Heart* **1999**, *82*, 547–554. [CrossRef] [PubMed]
10. Mao, Y.; Zhao, H.; Yu, C.; Yang, Y.; Ma, M.; Wang, Y.; Jiang, R.; Zhao, B.; Zheng, Z.; Jiang, C. Left Atrial Appendage Mechanical Dispersion Assessed by Speckle-Tracking Echocardiography as a Determinant of Left Atrial Appendage Blood Stasis in Patients With Atrial Fibrillation. *Front. Cardiovasc. Med.* **2022**, *9*, 905293. [CrossRef]
11. Musotto, G.; Monteleone, A.; Vella, D.; Di Leonardo, S.; Viola, A.; Pitarresi, G.; Zuccarello, B.; Pantano, A.; Cook, A.; Bosi, G.M.; et al. The Role of Patient-Specific Morphological Features of the Left Atrial Appendage on the Thromboembolic Risk Under Atrial Fibrillation. *Front. Cardiovasc. Med.* **2022**, *9*, 894187. [CrossRef]
12. Holzapfel, G.A. Similarities between Soft Biological Tissues and Rubber like Materials. In *Constitutive Models for Rubber IV*; Austrell, P.-E., Keri, L., Eds.; A.A. Balkema: Leiden, The Netherlands, 2005; Volume IV; pp. 607–617. ISBN 9781315140216.
13. de Oliveira-Santos, M.; Oliveira-Santos, E.; Gonçalves, L.; Silva Marques, J. Cardiovascular Three-Dimensional Printing in Non-Congenital Percutaneous Interventions. *Heart Lung Circ.* **2019**, *28*, 1525–1534. [CrossRef]
14. Otton, J.M.; Spina, R.; Sulas, R.; Subbiah, R.N.; Jacobs, N.; Muller, D.W.M.; Gunalingam, B. Left Atrial Appendage Closure Guided by Personalized 3D-Printed Cardiac Reconstruction. *JACC Cardiovasc. Interv.* **2015**, *8*, 1004–1006. [CrossRef] [PubMed]
15. Song, H.; Zhou, Q.; Zhang, L.; Deng, Q.; Wang, Y.; Hu, B.; Tan, T.; Chen, J.; Pan, Y.; He, F. Evaluating the Morphology of the Left Atrial Appendage by a Transesophageal Echocardiographic 3-Dimensional Printed Model. *Medicine* **2017**, *96*, e7865. [CrossRef]
16. Hachulla, A.L.; Noble, S.; Guglielmi, G.; Agulleiro, D.; Müller, H.; Vallée, J.P. 3D-Printed Heart Model to Guide LAA Closure: Useful in Clinical Practice? *Eur. Radiol.* **2019**, *29*, 251–258. [CrossRef]
17. Liu, P.; Liu, R.; Zhang, Y.; Liu, Y.; Tang, X.; Cheng, Y. The Value of 3D Printing Models of Left Atrial Appendage Using Real-Time 3D Transesophageal Echocardiographic Data in Left Atrial Appendage Occlusion: Applications toward an Era of Truly Personalized Medicine. *Cardiology* **2016**, *135*, 255–261. [CrossRef]
18. Hong, D.; Moon, S.; Cho, Y.; Oh, I.-Y.; Chun, E.J.; Kim, N. Rehearsal Simulation to Determine the Size of Device for Left Atrial Appendage Occlusion Using Patient-Specific 3D-Printed Phantoms. *Sci. Rep.* **2022**, *12*, 7746. [CrossRef] [PubMed]
19. Fan, Y.; Yang, F.; Cheung, G.S.-H.; Chan, A.K.-Y.; Wang, D.D.; Lam, Y.-Y.; Chow, M.C.-K.; Leong, M.C.-W.; Kam, K.K.-H.; So, K.C.-Y.; et al. Device Sizing Guided by Echocardiography-Based Three-Dimensional Printing Is Associated with Superior Outcome after Percutaneous Left Atrial Appendage Occlusion. *J. Am. Soc. Echocardiogr.* **2019**, *32*, 708–719.e1. [CrossRef]

20. Obasare, E.; Mainigi, S.K.; Morris, D.L.; Slipczuk, L.; Goykhman, I.; Friend, E.; Ziccardi, M.R.; Pressman, G.S. CT Based 3D Printing Is Superior to Transesophageal Echocardiography for Pre-Procedure Planning in Left Atrial Appendage Device Closure. *Int. J. Cardiovasc. Imaging* **2017**, *34*, 821–831. [CrossRef]
21. Hell, M.; Achenbach, S.; Yoo, I.; Franke, J.; Blachutzik, F.; Roether, J.; Graf, V.; Raaz-Schrauder, D.; Marwan, M.; Schlundt, C. 3D Printing for Sizing Left Atrial Appendage Closure Device: Head-to-Head Comparison with Computed Tomography and Transoesophageal Echocardiography. *EuroIntervention* **2017**, *13*, 1234–1241. [CrossRef] [PubMed]
22. Eng, M.H.; Wang, D.D.; Greenbaum, A.B.; Gheewala, N.; Kupsky, D.; Aka, T.; Song, T.; Kendall, B.J.; Wyman, J.; Myers, E.; et al. Prospective, Randomized Comparison of 3-dimensional Computed Tomography Guidance versus TEE Data for Left Atrial Appendage Occlusion (PRO3DLAAO). *Catheter. Cardiovasc. Interv.* **2018**, *92*, 401–407. [CrossRef] [PubMed]
23. Wang, D.D.; Eng, M.; Kupsky, D.; Myers, E.; Forbes, M.; Rahman, M.; Zaidan, M.; Parikh, S.; Wyman, J.; Pantelic, M.; et al. Application of 3-Dimensional Computed Tomographic Image Guidance to WATCHMAN Implantation and Impact on Early Operator Learning Curve. *JACC Cardiovasc. Interv.* **2016**, *9*, 2329–2340. [CrossRef]
24. Li, H.; Yao, Q.; Shen, B.; Shu, M.; Zhong, L.; Wang, X.; Song, Z. Application of 3D printing technology to left atrial appendage occlusion. *Int. J. Cardiol.* **2017**, *231*, 258–263. [CrossRef]
25. Valvez, S.; Oliveira-Santos, M.; Piedade, A.P.; Gonçalves, L.; Amaro, A.M. Planning the Closure Procedure: Additive Manufacture of a Left Atrial Appendage. In *Proceedings of the 10th Congress of the Portuguese Society of Biomechanics*; Martins Amaro, A., Roseiro, L., Messias, A.L., Gomes, B., Almeida, H., António Castro, M., Neto, M.A., de Fátima Paulino, M., Maranha, V., Eds.; Lecture Notes in Bioengineering; Springer Nature: Cham, Switzerland, 2023; pp. 89–95.
26. Cooke, M.N.; Fisher, J.P.; Dean, D.; Rimnac, C.; Mikos, A.G. Use of Stereolithography to Manufacture Critical-sized 3D Biodegradable Scaffolds for Bone Ingrowth. *J. Biomed. Mater. Res. B Appl. Biomater.* **2003**, *64B*, 65–69. [CrossRef]
27. Jain, A.; Choudhury, S.; Sundaresan, N.R.; Chatterjee, K. Essential Role of Anisotropy in Bioengineered Cardiac Tissue Models. *Adv. Biol.* **2024**, *8*, e2300197. [CrossRef] [PubMed]
28. Zhang, H.L.; Sinha, S.; Cameron, R.E.; Best, S.M. Mimicking Transmural Helical Cardiomyofibre Orientation Using Bouligand-like Pore Structures in Ice-Templated Collagen Scaffolds. *Polymers* **2023**, *15*, 4420. [CrossRef] [PubMed]
29. Dwyer, K.D.; Coulombe, K.L.K. Cardiac Mechanostructure: Using Mechanics and Anisotropy as Inspiration for Developing Epicardial Therapies in Treating Myocardial Infarction. *Bioact. Mater.* **2021**, *6*, 2198–2220. [CrossRef]
30. Wu, S.; Li, Y.; Zhang, C.; Tao, L.; Kuss, M.; Lim, J.Y.; Butcher, J.; Duan, B. Tri-Layered and Gel-Like Nanofibrous Scaffolds with Anisotropic Features for Engineering Heart Valve Leaflets. *Adv. Health Mater.* **2022**, *11*, 2200053. [CrossRef]
31. Lyra-Leite, D.M.; Andres, A.M.; Petersen, A.P.; Ariyasinghe, N.R.; Cho, N.; Lee, J.A.; Gottlieb, R.A.; Mccain, M.L. Mitochondrial Function in Engineered Cardiac Tissues Is Regulated by Extracellular Matrix Elasticity and Tissue Alignment. *Am. J. Physiol.-Heart Circ. Physiol.* **2017**, *313*, H757–H767. [CrossRef]
32. Lee, D.S.M.; Damrauer, S.M.; Levin, M.G. Genetics of Atrial Fibrillation. *Curr. Opin. Cardiol.* **2023**, *38*, 162–168. [CrossRef]
33. Jap, N.S.F.; Pearce, G.M.; Hellier, A.K.; Russell, N.; Parr, W.C.; Walsh, W.R. The Effect of Raster Orientation on the Static and Fatigue Properties of Filament Deposited ABS Polymer. *Int. J. Fatigue* **2019**, *124*, 328–337. [CrossRef]
34. Naveed, N. Investigating the Material Properties and Microstructural Changes of Fused Filament Fabricated PLA and Tough-PLA Parts. *Polymers* **2021**, *13*, 1487. [CrossRef]
35. Lanzotti, A.; Grasso, M.; Staiano, G.; Martorelli, M. The Impact of Process Parameters on Mechanical Properties of Parts Fabricated in PLA with an Open-Source 3-D Printer. *Rapid Prototyp. J.* **2015**, *21*, 604–617. [CrossRef]
36. Palanisamy, C.; Aaron Tay Hong Kiat, H.K. Experimental Study and Modeling the Tensile Strength of 3D-printed Aluminium Polylactic Acid (PLA) Parts Using Artificial Neural Networks. *F1000Res* **2021**, *10*, 1286. [CrossRef]
37. Brandt-Wunderlich, C.; Schmidt, W.; Grabow, N.; Stiehm, M.; Siewert, S.; Andresen, R.; Schmitz, K.-P. Support Function of Self-Expanding Nitinol Stents—Are Radial Resistive Force and Crush Resistance Comparable? *Curr. Dir. Biomed. Eng.* **2019**, *5*, 465–467. [CrossRef]
38. Toong, D.W.Y.; Ng, J.C.K.; Huang, Y.; Wong, P.E.H.; Leo, H.L.; Venkatraman, S.S.; Ang, H.Y. Bioresorbable Metals in Cardiovascular Stents: Material Insights and Progress. *Materialia* **2020**, *12*, 100727. [CrossRef]
39. *ISO 527*; Plastics—Determination of Tensile Properties—Part 1: General Principles. Blandonnet 8 CH-1214 Vernier: Geneva, Switzerland, 2019.
40. FUNG, Y.C. Structure and Stress-Strain Relationship of Soft Tissues. *Am. Zool.* **1984**, *24*, 13–22. [CrossRef]
41. Mahmood, S.; Qureshi, A.J.; Goh, K.L.; Talamona, D. Tensile Strength of Partially Filled FFF Printed Parts: Experimental Results. *Rapid Prototyp. J.* **2017**, *23*, 122–128. [CrossRef]
42. Dong, Y.; Milentis, J.; Pramanik, A. Additive Manufacturing of Mechanical Testing Samples Based on Virgin Poly (Lactic Acid) (PLA) and PLA/Wood Fibre Composites. *Adv. Manuf.* **2018**, *6*, 71–82. [CrossRef]

43. Taşdemir, V. Investigation of the Effects of the Number of Shells, Raster Angle, Extrusion Ratio, and Path Width on Printed Polylactic Acid Parts with Fused Deposition Modeling 3D Printer. *J. Mater. Eng. Perform.* **2024**, *33*, 11888–11898. [CrossRef]
44. Kowalska, N.; Szczygieł, P.; Skrzyniarz, M.; Błasiak, S. Effect of Shells Number and Machining on Selected Properties of 3D-Printed PLA Samples. *Polimery* **2024**, *69*, 186–190. [CrossRef]
45. Zhong, F.; Xu, Y.; Zhao, H.; Lu, L. As-Continuous-As-Possible Extrusion-Based Fabrication of Surface Models. *ACM Trans. Graph.* **2023**, *42*, 1–16. [CrossRef]
46. Szot, W.; Rudnik, M. Effect of the Number of Shells on Selected Mechanical Properties of Parts Manufactured by FDM/FFF Technology. *Adv. Mater. Sci.* **2024**, *24*, 86–103. [CrossRef]
47. Shulga, E.; Karamov, R.; Sergeichev, I.S.; Konev, S.D.; Shurygina, L.I.; Akhatov, I.S.; Shandakov, S.D.; Nasibulin, A.G. Fused Filament Fabricated Polypropylene Composite Reinforced by Aligned Glass Fibers. *Materials* **2020**, *13*, 3442. [CrossRef] [PubMed]
48. Chicos, L.-A.; Pop, M.A.; Zaharia, S.-M.; Lancea, C.; Buican, G.R.; Pascariu, I.S.; Stamate, V.-M. Infill Density Influence on Mechanical and Thermal Properties of Short Carbon Fiber-Reinforced Polyamide Composites Manufactured by FFF Process. *Materials* **2022**, *15*, 3706. [CrossRef]
49. Zabihollah, A.; Bardowell, S.; Does, B.; Vuddandam, R. Experimental Determination of Tensile Stress Concentration Factors of 3D-Printed PLA Polymers: Fillet Radius, Hole Diameter, and Infill Density. *Adv. Mat. Res.* **2024**, *1180*, 13–22. [CrossRef]
50. Bai, Y.; Gao, J.; Huang, C.; Li, Y. Mechanical Properties and Energy Absorption of Integrated AlSi10Mg Shell Structures with BCC Lattice Infill. *Chin. J. Mech. Eng.* **2023**, *36*, 143. [CrossRef]
51. Khaliq, J.; Gurrapu, D.R.; Elfakhri, F. Effects of Infill Line Multiplier and Patterns on Mechanical Properties of Lightweight and Resilient Hollow Section Products Manufactured Using Fused Filament Fabrication. *Polymers* **2023**, *15*, 2585. [CrossRef] [PubMed]
52. Javani, S.; Gordon, M.; Azadani, A.N. Biomechanical Properties and Microstructure of Heart Chambers: A Paired Comparison Study in an Ovine Model. *Ann. Biomed. Eng.* **2016**, *44*, 3266–3283. [CrossRef]
53. Islam, M.R.; Virag, J.; Oyen, M.L. Micromechanical Poroelastic and Viscoelastic Properties of Ex-Vivo Soft Tissues. *J. Biomech.* **2020**, *113*, 110090. [CrossRef]
54. Martonová, D.; Alkassar, M.; Seufert, J.; Holz, D.; Dương, M.T.; Reischl, B.; Friedrich, O.; Leyendecker, S. Passive Mechanical Properties in Healthy and Infarcted Rat Left Ventricle Characterised via a Mixture Model. *J. Mech. Behav. Biomed. Mater.* **2021**, *119*, 104430. [CrossRef] [PubMed]
55. Chang, Z.; Zhang, J.; Liu, Y.; Gao, H.; Xu, G.-K. New Mechanical Markers for Tracking the Progression of Myocardial Infarction. *Nano Lett.* **2023**, *23*, 7350–7357. [CrossRef]
56. Goh, G.D.; Sing, S.L.; Lim, Y.F.; Thong, J.L.J.; Peh, Z.K.; Mogali, S.R.; Yeong, W.Y. Machine Learning for 3D Printed Multi-Materials Tissue-Mimicking Anatomical Models. *Mater. Des.* **2021**, *211*, 110125. [CrossRef]
57. Bregasi, A.; Freeman, J.V.; Curtis, J.P.; Akar, J.G.; Ortiz-Leon, X.A.; Maia, J.H.; Higgins, A.Y.; Matthews, R.V.; Sinusas, A.J.; McNamara, R.L.; et al. Abnormal Left Atrial Body Stiffness Is Predicted by Appendage Size: Impact of Appendage Occlusion on Left Atrial Mechanics Assessed by Pressure-Volume Analysis. *Am. J. Physiol.-Heart Circ. Physiol.* **2022**, *323*, H559–H568. [CrossRef] [PubMed]

MDPI AG
Grosspeteranlage 5
4052 Basel
Switzerland
Tel.: +41 61 683 77 34

Polymers Editorial Office
E-mail: polymers@mdpi.com
www.mdpi.com/journal/polymers

www.ingramcontent.com/pod-product-compliance
Lightning Source LLC
Chambersburg PA
CBHW062100220926
51850CB00010B/2832
* 9 7 8 3 7 2 5 8 6 9 5 6 5 *